DNA Markers in Plant Improvement

DNA Markers in Plant Improvement

Editor
S.K. Mangal

GENE TECH

2012

Gene-Tech Books
New Delhi - 110 002

ISBN 81-89729-28-4

Published by : **GENE-TECH BOOKS**
4762-63/23, Ansari Road, Darya Ganj,
NEW DELHI - 110 002
Phone: 41562849
e-mail: genetechbooks@yahoo.co.in

Printed at : **Tarun Offset Printers**
Delhi

PRINTED IN INDIA

Preface

The progress made in DNA marker technology has been tremendous and exciting. DNA markers have provided valuable tools in various analyses ranging from phylogenetic analysis to the positional cloning of genes. The development of high-density molecular maps which has been facilitated by PCR-based markers, have made the mapping and tagging of almost any trait possible. Marker-assisted selection has the potential to deploy favourable gene combinations for disease control.

Comparative studies between incompatible species using DNA markers has resulted in synteny maps which are useful not only in predicting genome organization and evolution but also have practical application in plant breeding. DNA marker technology has found application in fingerprinting genotypes, in determining seed purity, in systematic sampling of germplasm, and in phylogenetic analysis.

With the advent of DNA marker technology, several types of DNA markers and molecular breeding strategies are now available to plant breeders and geneticists, helping them to overcome many of the problems faced during conventional breeding. This book elaborates on the applications of DNA marker technology in the genetic analysis and improvement of plants.

This book discusses the use of DNA marker technology for the genetic improvement of plants. Its' contents will be immensely useful for teachers, students and researchers of varied fields of life sciences.

Editor

Contents

1

Plant Biotechnology

Today, biotechnology is being used as a tool to give plants new traits that benefit agricultural production, the environment, and human nutrition and health. The goal of plant breeding is to combine desirable traits from different varieties of plants to produce plants of superior quality. This approach to improving crop production has been very successful over the years.

For example, it would be beneficial to cross a tomato plant that bears sweeter fruit with one that exhibits increased disease resistance. To do this, it takes many years of crossing and backcrossing generations of plants to obtain the desired trait. Along the way, undesirable traits may be manifested in the plants because there is no way to select for one trait without affecting others. Another limitation of traditional plant selection is that breeding is restricted to plants that can sexually mate.

Advances in scientific discovery and laboratory techniques during the last half of the twentieth century led to the ability to manipulate the deoxyribonucleic acid (DNA) of organisms, which accelerated the process of plant improvement through the use of biotechnology.

Genes and the Genome

Plants are made of millions of cells all working together. Every cell of a plant has a complete "instruction manual" or genome that is inherited from the parents of the plant as a combination of their genomes. Genes are found within the genome and serve as the "words" of the instruction manual. When a cell reads a word, or in scientific terms "expresses a gene," a specific protein is produced. Proteins give an individual cell, and therefore the plant, its form and function. Genes (words) are written using the four-letter alphabet A, C, G, T. The letters are abbreviations for four chemicals called bases, which together make up DNA. DNA is universal in nature, meaning that the four chemical bases of DNA are the same in all living organisms. Consequently, a gene from one organism can function in any other organism.

The ability to move genes into plants from other organisms, thereby producing new proteins in the plant, has resulted in significant achievements in plant biotechnology that were not possible using traditional breeding practices. To genetically modify a plant, the thousands of bases of DNA comprising an individual gene are transferred into an individual plant cell where the new gene becomes a permanent part of the cell's genome. This process makes the resulting plant "transgenic." Transfer of DNA into plant cells is done using various "transformation" techniques that are the result of discoveries in basic science.

One method to transfer DNA into plants takes advantage of a system found in nature. The bacterium that causes "crown gall tumours" injects its DNA into a plant genome, forcing the plant to create a suitable environment for the bacterium to live. After discovering this process, scientists were able to "disarm" the bacterium, put new genes into it, and use the bacterium to harmlessly insert the desired genes into the plant genome.

In the "biolistic" or "gene gun" method, microscopic gold beads are coated with the gene of interest and shot into the plant cell with a burst of helium. Once inside the cell, the gene comes off the bead and integrates into the cell's genome. It was also discovered that plant cells could be "electroporated" or mixed with a gene and "shocked" with a pulse of electricity, causing holes to form in the cell through which the DNA could flow. The cell is subsequently able to repair the holes and the gene becomes a part of the plant genome.

When using these methods, new genes are successfully introduced into only a small percentage of the cells, so scientists must be able to "pick out" or "select" the transformed cells before proceeding. This is often done by concurrently introducing an additional gene into the cell that will make it resistant to an antibiotic. A cell that survives antibiotic treatment will most likely have received the gene of interest as well; that cell is subsequently used to propagate the new plant. There is a concern that the gene giving antibiotic resistance could naturally be transferred to bacteria once the transgenic plant is in the wild, making bacteria resistant to antibiotics that are used to fight human infection. Scientists are currently devising ways to select for transformed cells that will alleviate this issue.

Traits in Plants

Changes made to plants through the use of biotechnology can be categorised into the three broad areas of input, output, and value-added traits. Examples of each are described below.

Input traits

An "input" trait helps producers by lowering the cost of production, improving crop yields, and reducing the level of chemicals required for the control of insects, diseases, and weeds.

Input traits that are commercially available or being tested in plants:

- — Resistance to destruction by insects
- — Tolerance to broad-spectrum herbicides
- — Resistance to diseases caused by viruses, bacteria, fungi, and worms
- — Protection from environmental stresses such as heat, cold, drought, and high salt concentration

Plants with "input traits" that are commercially available include:

- — Roundup Ready® soybean, canola, and corn: resistant to treatment with Roundup herbicide that may result in more effective weed control with less tillage, and/or decreased use of other, more harmful herbicides
- — YieldGard® corn and Bollgard® cotton: express an insecticidal protein that is not toxic to animals or humans which protects the plant from damage caused by the European corn borer, tobacco budworm, and bollworm
- — Destiny III® and Liberator III® squash: resistant to some viruses that destroy squash

Output Traits

An "output" trait helps consumers by enhancing the quality of the food and fiber products they use. Output traits that consumers may one day be able to take advantage of:

— Nutritionally enhanced foods that contain more starch or protein, more vitamins, more anti-oxidants (to reduce the risk of certain cancers), and fewer trans-fatty acids (to lower the risk of heart disease)

— Foods with improved taste, increased shelf-life, and better ripening characteristics

— Trees that make it possible to produce paper with less environmental damage

— Nicotine-free tobacco

— Ornamental flowers with new colours, fragrances, and increased longevity

Plants may become available with "output traits" including:

— High laurate canola and high oleic soybean having altered oil content to be used primarily in industrial oils and fluids rather than food

— High-starch potatoes that take up less oil when frying

— Longer shelf-life bananas, peppers, pineapples, strawberries, and tomatoes

— Soybeans with higher levels of isoflavones; compounds that may be beneficial in reducing some cancers and heart disease

— Plants that produce vaccines and pharmaceuticals for treatment of human diseases

— Corn with improved digestibility and more nutrients providing livestock with better feed

Value-added Traits

Genes are being placed into plants that completely

change the way they are used. Plants may be used as "manufacturing facilities" to inexpensively produce large quantities of materials including:

— Therapeutic proteins for disease treatment and vaccination

— Textile fibers

— Biodegradable plastics

— Oils for use in paints, detergents, and lubricants

Plants are being produced with entirely new functions that enable them to do things such as:

— Detect and/or dispose of environmental contaminants like mercury, lead, and petroleum products

Genetically Modified Plants

The list of plants and plant-derived products made as a result of modern biotechnology is ever- increasing. Many transgenic plants, such as herbicide resistant soybeans, have been widely adopted by producers signifying their satisfaction, while other products, such as the delayed softening "FlavrSavr" tomato, are no longer on the market. Some of the potential benefits from using transgenic plants include:

— Reduced crop production costs and increased yields

— Healthier, more nutritious foods

— Reduced environmental impact from farming and industry

— Increased food availability for underdeveloped countries

Potential risks associated with transgenic plants include:

— Introduction of allergenic or otherwise harmful proteins into foods

— Transfer of transgenic properties to viruses, bacteria, or other plants

— Detrimental effects on non-target species and the environment

Safety, Regulation, and Labelling

In the United States, at the Federal level, the Food and Drug Administration, the Environmental Protection Agency, and the Department of Agriculture extensively review products of biotechnology to ensure that they are safe for public use and the environment. GM foods require labelling only if they differ significantly in safety, composition, or nutritional content when compared to their non-GM counterparts. Additionally, the FDA requires a GM food to be labeled if it contains a known allergen unless data have shown that there is no allergy risk.

Organic standards reflect a "zero tolerance" policy concerning transgenic products and organisms. Organic food producers are taking precautions to minimise the risk of unintentional contamination of their products with transgenic ones.

The Canadian Food Inspection Agency, Health Canada, and Environment Canada strictly regulate agricultural biotechnology products. They currently require GM foods to be labeled if they differ significantly in composition or nutritional value and support a voluntary labelling policy for others.

The acceptance of GM crops by the European Union has been more reserved. However, recent statements made by European Union officials suggest that their position may be changing as they are calling for their policies regarding GM crops to be based on scientific principles rather than on public opinion and misconceptions. Europe currently favours labelling of all GM foods and a system that would allow for "identity preserved" processing in which foods would be guaranteed to contain no genetically modified products.

Genetic Improvement of Food Crops

The end of a year, decade, century, or, as now, of a millennium, always offers an opportunity to reflect on human activity in a particular discipline and to formulate a future strategy. Researchers constantly examine past occurrences in order to learn lessons that could help in the acquisition of new knowledge or for the further development of appropriate technology ensuing from it. Of course, science and technology are not isolated in the world, so researchers are expected to act according to the changing global society in which they live.

This behaviour could be seen as the major challenge of crop biotechnology for the next millennium, i.e., to consider the social actors in the research agenda and work. In other words, market forces, user demands, and public views cannot be ignored when addressing basic and strategic research issues because these factors shape scientific investigations and technology or product development.

It would be inappropriate to attempt to provide an outlook beyond this time-span because of the ever accelerating progress in this field. For example 15 years ago, plant biotechnology comprised only a few applications of tissue culture, recombinant DNA technology and monoclonal antibodies. Today, transformation, and marker-aided selection and breeding are just a few of the examples of the applications of biotechnology in crop improvement.

Writing about biotechnology for crop improvement in the next millennium does not appear to be an easy task owing to the rapid progress in this field. Within the last 100 years the world has seen the rise of genetics as a scientific discipline, the finding of DNA as the hereditary material, the elucidation of the double helix structure of the DNA molecule, the cracking of the genetic code, the ability to isolate genes, and the application of DNA recombinant techniques.

Methods of crop improvement have also changed dramatically throughout this century. Mass and pure line selection in landraces, consisting of genotype mixtures, were the popular breeding techniques until the 1930s for most crops. In the 1930s maize breeders started the commercial development of double cross hybrids that was followed by the extensive utilisation of single crop hybrids since the 1960s. Pedigree-, bulk-, backcross- and other selection methods were also developed especially for self-pollinating crop species.

Such scientific advances in plant breeding led to the so-called 'Green Revolution', one of the greatest achievements to feed the world in the years of the Cold War. Owing to this agricultural betterment, cereal production, which accounts for more than 50% of the total energy intake of the world's poor, kept in pace with the high average population growth rate of 1.8% since 1950.

Today, 370 kg of cereals per person are harvested as compared to only 275 kg in the 1950s; i.e., in excess of 33% per capita gain. Similar progress in other food crops resulted in 20% per capita gains since the early 1960s, according to FAO. There are 150 million fewer hungry people in the world today than 40 years ago, though there are twice as many human beings. Despite this splendid progress in crop productivity, even greater progress must be made in order to feed an additional two billion people by the early part of the 21st century.

Around 800 million people are hungry today and another 185 million pre-school children are still malnourished owing to lack of food and water, or disease. Hence as suggested by the Nobel Peace Laureate, Norman Bourlag, new biotechniques, in addition to conventional plant breeding, are needed to boost yields of the crops that feed the world. Careful choice of such biotechniques as well as a realistic assessment of their potential in crop improvement are needed to avoid not only the criticism of the anti-science lobbyists but also the permanent distrust of pragmatic traditional breeders.

For example, a World Bank panel recently released for discussion a well based report concerning bioengineering of crops. Indeed, plant biotechnology has been regarded as a priority area for technology transfer, because genetically modified food, feed, and fibre are of vital concern to the developing world. Therefore, the rich industrialised world should share their biotechniques and avoid policies that do not allow the progress of agriculture in poor, non-industrialised parts of the world, where this economic activity still provides 60 to 80% employment and 50% of national income.

Such support will assist the developing world towards food self-reliance, which will be very important to avoid hunger and keep peace in many regions of the tropics, where the agricultural sector remains the most important basis for economic growth. Furthermore, a wealthy society provides high living standards to its citizens. Tissue culture was developed in the 1950s and became popular in the 1960s. Today, micropropagation and in vitro conservation are standard techniques in most important crops, especially those with vegetative propagation.

At the beginning of the 1980s genetic engineering of plant remained a promise of the future, although gene transfer had already been achieved earlier in a bacterium. The first transgenic plant, a tobacco accession resistant to an antibiotic, was reported in 1983. Transgenic crops with herbicide, virus or insect resistance, delayed fruit ripening, male sterility, and new chemical composition have been released to the market in this decade.

In 1996, there were about 3 million ha of transgenic crops grown in the world (mainly in North America) whereas in excess of 34 million ha (a 12-fold addition) of transgenic crops will be harvested this year in North America, Argentina, China, and South Africa among other countries. Argentina is the leading developing country with an excess of 4 million ha of transgenic herbicide-resistant soybean. There are 4.4 million ha of transgenic corn (14% of total acreage), 5 million ha of transgenic soybean (20%), and 1.6 million ha of transgenic canola (42%) grown only in North America. It has been calculated that in 1998 US farmers are growing over 50% of their cotton fields with transgenic seeds, the largest percentage for any crop ever. Trees are the next target in the agenda of genetic engineering.

Allozymes were available as the first biochemical genetic markers in the 1960s. Population geneticists took advantage of such marker system for their early research. In the 1970s, restriction fragment length polymorphisms (RFLP) and Southern blotting were added to the tool box of the geneticists. Taq polymerase was found in the 1980s, and the polymerase chain reaction (PCR) developed shortly afterwards. Since then, marker-aided analysis based on PCR have become routine in plant genetic research and marker systems have shown their potential in plant breeding.

Furthermore, new single nucleotide polymorphic markers based on high density DNA arrays, a technique known as 'gene chips', have recently been developed. With 'gene chips', DNA belonging to thousand of genes can be arranged in small matrices (or chips) and probed with labeled cDNA from a tissue of choice. DNA chip technology uses microscopic arrays (or micro-arrays) of molecules immobilised on solid surfaces for biochemical analysis. An electronic device connected to a computer may read this information, which will facilitate marker-assisted selection in crop breeding.

In short, since Mendel's work on peas, there have been five eras in genetic marker evolution: morphology and cytology in early genetics, protein and allozyme electrophoresis in the pre-recombinant DNA time, RFLP and minisatellites in the pre-PCR age, random amplified polymorphic DNA, microsatellites, expressed sequence tags, sequence tagged sites, and amplified fragment length polymorphism in the oligoscene period, and complete DNA sequences with known or unknown function as well as complete protein catalogs in the current computer robotic cyber genetics generation. The driving force for such a development has been the scientific interest of human beings to understand and manipulate the inheritance of their own characters.

Improvements in Plant Transgenics and Genomics

The advances in plant transgenics and genomics described above have not been isolated from society. Some of these achievements have been acclaimed by end-users whereas other accomplishments, e.g. release of genetically modified organisms (GMO), are being attacked, not only in words but also in deeds, by political activists. Some of these educated middle-class campaigners are expressing in this way their rampant 'eco-paranoia', while others hide their real agenda to manipulate the fashionable ecological movement. This controversy has attracted the attention of non-scientific partizans to each side.

There have been negative comments about transgenic plants by a crown prince and contrasting positive comments by a former president, both of whom may not have the required technical knowledge to assess the potential of biotechnology for crop improvement. Irrespective of this ideological dispute and ensuing democratic disagreements, biotechnology products will be accepted by people who support scientific-based progress, in a similar way that new cultivars or innovative crop husbandry techniques have previously become integral parts of farming systems elsewhere. However, without end-user's consent, the impact of a new technology in the society will be small or nil.

Scientific honesty seems to the best policy to convince people about the advantages of biotechnology for crop improvement. What to do? Scientists, farmers, consumers, and policy-makers should objectively assess the potential hazards of crop biotechnology in farming and food systems regarding the current situation and the likelihood that such hazards may occur. For example scientists should explain to the people that gene recombination (or reassortment) already occurs in nature. However, the ecological success of viable recombinants after gene reassortment is unpredictable owing to the high fitness of current isolates. For this reason, more scientific research will be needed to identify unpredictable risks and the chances of their occurrence.

The need for profit, as in any other business, has attracted the interest of the private sector to defend their investments in crop biotechnology with patents, intellectual property rights, and new protection methods, e.g. 'terminator' technology that inhibits germination of self-pollinated seeds. This technology protection system prevents farmers from saving seeds from their harvest for further utilisation as next season planting propagules. Three genes, each with a specific promoter, are inserted into the 'terminator' plant.

One of the genes (e.g. CRE/LOX system from bacteriophages) produces a recombinase that removes a spacer between the gene producing, for example, a ribosomal inhibitor protein and its promoter such as late embryonic abundance, which only becomes active during the late stages of embryo development. This spacer with specific recognition sites blocks the gene (for the ribosomal inhibitor protein) from being activated. Another gene (e.g. tetracycline repressor system) produces a repressor that keeps off the recombinase gene until an outside stimulus is applied to the 'terminator' plant, e.g. a chemical such as the tetracycline, or temperature and osmotic shocks.

The United States Department of Agriculture (USDA) and a cotton seed enterprise jointly acquired a patent for this concept (U.S. patent 5,723,765). Two months after this patent was announced, one of the leading agro-chemical transnationals bought the cotton seed company, although one of its officers said that it may take many years before this 'terminator gene' idea becomes a proven technology in the seed industry.

Strategic alliances, joint ventures, research partnerships, new investments, company mergers, cross-ownerships, and take-overs in the seed and agro-chemical business have also been in the news in recent months. Likewise, some leading scientists are leaving their academic appointments to join the new private enterprises in plant biotechnology.

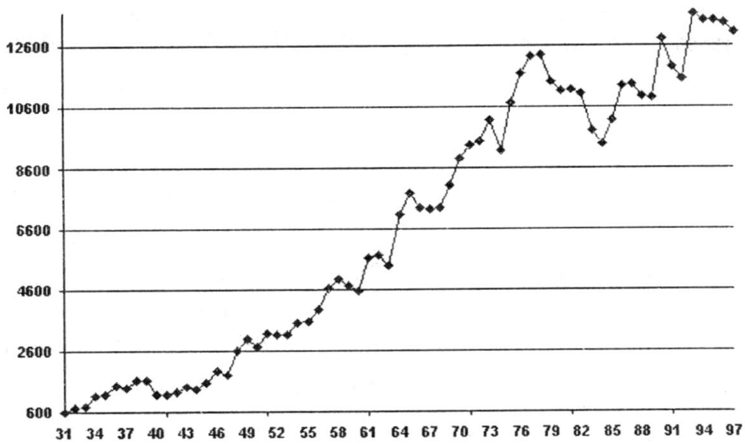

Figure 1. Number of publications indexed in Plant Breeding Abstracts since its publication in early 1930s

These events are happening because the private sector wants to use biotechnology to accelerate its growth in agri-business in the short-term. Nonetheless, funds to support basic and strategic research by public researchers are needed for a long-term sustainable transfer of public goods (both knowledge and technology) to the private sector or other users.

Bioinformatics

Another important factor in the successes of the genetic improvement of crops was the development of fast and more reliable computers, which allowed easier management and analysis of data as well as publication of scientific reports. The impact of the informatic revolution in crop improvement can be partially assessed by counting the number of publications indexed in Plant Breeding Abstracts. There was ca. 22-fold increase of publications in the 1930-1997 period (Figure 1). It was in the 1970s that indexed publications in plant breeding exceeded 10,000 per year. More publications and easy means for retrieving this information accounted for such growth of knowledge dissemination in plant genetics and breeding.

Today, rapid information exchange has been facilitated with electronic mail and access to the internet to read electronic publications such as this journal. Nowadays, information technology and DNA science are beginning to fuse into a single operation. Computers are deciphering, and organising the huge genetic information that may become "the raw resource of the emerging biotech economy" in the next century. Scientists working in the new field of "bioinformatics" are developing biological data banks to download the genetic information accumulated during millions of years of life evolution, and perhaps reconstruct some of the living organisms of the natural world.

Plant Genomics

This new term, defined by the development of biotechnology, refers to the investigations of whole genomes by integrating genetics with informatics and automated systems. Genomic research aims to elucidate the structure, function and evolution of past and present genomes. Some of the most dynamic fields concerning agriculture are the sequencing of plant genomes, comparative mapping across species with genetic markers, and objective assisted breeding after identifying candidate genes or chromosome regions for further manipulations. As a result of genomics, the concept of gene pools has been enlarged to include transgenes and native exotic gene pools that are becoming available through comparative analysis of plant biological repertoires. Understanding the biological traits of one species may enhance the ability to achieve high productivity or better product quality in another organism.

DNA markers and gene sequencing provides quantitative means to determine the extent of genetic diversity and to establish objective phylogenetic relationships among organisms. 'Gene chips' and transposon tagging will provide new dimensions for investigating gene expression. Molecular biologists will study not only individual genes but how circuits of interacting genes in different pathways control the spectrum of genetic diversity in any crop species. For example, more information will be available on why plant resistance genes are clustered together, or what candidate genes should be considered when manipulating quantitative trait loci (QTL) for crop improvement.

The aims of applied plant science research for agriculture are to enhance crop yields, improve food quality, and preserve the environment where human beings and other organisms live. The best way for conservation of plant biodiversity and its environment, would be to achieve high crop productivity per unit area. There has been a significant yield improvement owing to enhanced crop husbandry, but in the next years progress will be achieved by changing plants that could be more suitable to sustainable and environmentally friendly farming systems.

Agro-chemical corporations are developing pest and disease resistant transgenic crops to avoid pollution with pesticides in the farming system. Furthermore, food quality will become more important than crop productivity in a wealthy society. Consumers will prefer transgenic crops if they have the desired characteristics.

In the next decades meiotic-based breeding will still generate cultivars for farmers. Genetic improvement through biotechnology needs conventional breeding because (1) the elite cultivars will be the parents of the next generation of improved genotypes, (2) field testing across locations or cropping systems and over years will be needed to determine the best selections due to the genotype-by-environment interaction.

As stated by Briggs, "transgenes must be viewed as improvements rather than replacements for elite germplasm". Indeed, genetic engineering may provide a means to add value by introducing synthetic or natural genes that enhance crop quality and yield, as well as protect the plant against pest and diseases. Farmers will pay more for transgenic crop propagules if they obtain extra-income after adopting biotech-derived products. For example, seeds of insect resistant transgenic crops will be more expensive than those of available cultivars but the farmer will not need to apply pesticides in their transgenic fields. Of course, patents make transgenic seeds more expensive but also farmer's benefits may be higher.

DNA Banking and Virtual Plant Breeding

The sequencing of crop genomes opened new frontiers in conservation of plant biodiversity and its genetic enhancement. The advances in gene isolation and sequencing in many plant species allows to envisage that within a few years, gene-bank curators may replace their large cold stores of seeds with crop DNA sequences that will be electronically stored. The characterisation of plant genomes will ultimately create a true gene bank, which should possess a large and accessible gene inventory of today's non-characterised crop gene pools. Of course, seed banks of comprehensively investigated stocks should remain because geneticists and plant breeders, the main users of gene banks, will need this germplasm for their work.

Genomics may accelerate the utilisation of candidate genes available at these gene banks through transformation without barriers across plant species or other living kingdoms. Nonetheless, genetic engineering should be seen as one of the methods of plant breeding that permits the direct alteration and re-building of a crop population. "Shutting-off" genes coding for undesired characteristics may be another application of transgenics in crop improvement.

Plant breeders will change their modus operandi with the development of objective marker-assisted introgression and selection methods. Backcross breeding will be shortened by eliminating undesired chromosome segments (also known as linkage drags) of the donor parent or selecting for more chromosome regions of the recurrent parent. Parents of elite crosses may be chosen based on a combination of DNA markers and phenotypic assessment in a selection index, such as best linear unbiased predictors. To achieve success in these endeavours, cheap, easy, decentralised, and rapid diagnostic marker procedures are required.

There are many areas of basic and strategic research in plant breeding and genetics that are being facilitated by marker-aided analysis. With molecular markers, plant biologists are reviewing crop evolution and gathering new knowledge. Such information should be incorporated into genetic enhancement programmes, especially those with an evolutionary breeding scheme. Likewise, plant ideotypes for each crop should drive the work of plant breeders. Specific plant morphotypes have been defined in rice and wheat based on accumulated knowledge of crop physiology and crop protection. The needed characteristics required to develop improved plant prototypes ensuing from such a 'virtual breeding' approach may be available in gene banks of the crop or in those of other species. Otherwise, breeders may obtain novel transgenes to develop the required ideotype.

Nowadays, the finding of new genes that add value to agricultural products seems to be very important in the private agri-business. Unique gene databases are being assembled by the industry with the massive amount of data generated by genomics research. A new term 'biosource' was coined recently to refer to a fast and effective licensed technology of pinpointing genes. With this method, a 'benign' virus infects a plant with a specific gene that allows researchers to observe directly its phenotype. Biosource replaces the standard time-consuming approach of first mapping a gene to subsequently determine its exact function. Gene identification in DNA libraries coupled with biosource technology and an enhanced ability to put genes into plants will be routine for improving crops in the next decade.

Genomics may provide a means for the elucidation of important functions that are essential for crop adaptedness. Regions of the world should be mapped by combining data of geographical information systems, crop performance, and genome characterisation in each environment. In this way, plant breeders can develop new cultivars with the appropriate genes that improve fitness of the promising selections. Fine-tuning plant responses to distinct environments may enhance crop productivity.

Development of cultivars with a wide range of adaptation will allow farming in marginal lands. Likewise, research advances in gene regulation, especially those processses concerning plant development patterns, will help breeders to fit genotypes in specific environments. Photoperiod insensitivity, flowering initiation, vernalisation, cold acclimation, heat tolerance, host response to parasites and predators, are some of the characteristics in which advanced knowledge may be acquired by combining molecular biology, plant physiology and anatomy, crop protection, and genomics. Multidisciplinary co-operation among researchers will provide the required holistic approach to facilitate research progress in these subjects.

Pharming

Growth of cities in the developed world has already replaced farmland with shopping malls, parking lots, and housing developments. Peri-urban agriculture and home gardening are also becoming very important for national food security in the developing world as a result of rapid urban expansion. Hence, new cultivars will be needed to fit into intensive production systems, which may provide the food required to satisfy urban world demands of the next century. Specific plant architecture, tolerance to urban pollution, efficient nutrient uptake, and crop acclimatisation to new substrates for growing are, among others, the plant characteristics required for this kind of agriculture.

Genes controlling these characteristics may be available in gene banks for further cross breeding, which can be assisted by genomics. Peri-urban and home garden "farmers" will have to adapt to new demands from emerging urban populations with higher income. These consumers may request a more varied diet. For example, food crops with low fats, and high in specific amino acids may be needed to satisfy people who wish to change their eating habits. If genes controlling these characteristics do not exist in a specific crop pool they may be incorporated into the breeding pool using transgenics.

Tissue culture of certain parts of the plant may provide a means to achieve success in this endeavour. For example, edible portions of fruit crops could be grown in vitro. A steady and cheap supply of these edible plant parts will be required in this new agri-business. It will take some time before such a process can be scaled-up for commercial output. Nonetheless, a patent was submitted in 1991 by a Californian biotech company for producing a vanilla extract through cell culture. Of course, this technique will not replace farming as we know it today. This biotechnique, as well as other new farming methods, offers a means for new ways of producing food, feed or fibre.

Often plants provide the raw materials for agro-industry, and not only for food or fibre processing. Active ingredients of plants have been transformed into commercial products such as medicines, solvents, dyes, and non-cooking oils for many years. Hence, it would not be surprising to see, in few years from now, entire farms without food crops but growing transgenic plants to produce new products, e.g. edible plastic from peas or plant oils to manufacture hydraulic fluids and nylon. This new rural activity may result in important changes in the national economic sector.

'Pharming' has been added to the dictionary to indicate a new kind of system to obtain medicines. For example, oral vaccines appear to be a convenient delivery system for vaccination throughout the world. Biotechnology has been used to engineer plants that contain a gene derived from a human pathogen. An antigenic protein encoded by this foreign DNA can accumulate in the resultant plant tissues. Results from pre-clinical trials showed that antigenic proteins harvested from transgenic plants were able to keep the immunogenic properties if purified.

These antigenic proteins caused the production of specific antibodies in injected mice. Mice, which ate these transgenic plant tissues, also showed also a mucosal immune response. Arakawa et al. recently demonstrated the ability of transgenic food crops to induce protective immunity in mice against a bacterial enterotoxin such as cholera toxin B subunit pentamer with affinity for GMI-ganglioside. Also, potato tubers have been used successfully as a biofactory for high-level output of a recombinant single chain antibody.

Risk assessment of Transgenic Crops

Lack of scientific data, non-scientific partizan views, uncertainty of potential risks, and ignorance confound rational discussion concerning the release of GMO. The issue of releasing genetically modified plants (GMP) into the farming system has become particularly agitated by lobbyist groups in Europe despite widespread cultivation of such crops in North America and elsewhere. Scientists must realise that the general public are concerned that an uncautious approach to the manipulation and cultivation of transgenic crops may affect biodiversity and its sustainable utilisation in the farming system, e.g. loss of variability and viability. People also want that their views about applications of biotechnology for improving agriculture are listened irrespective of their knowledge in the subject.

Moreover, farmers are afraid that negative propaganda jeopardizes the public image of their products. Scientists and policy markers should not forget that people's acceptability is the most important component of the general public assessment of risk, which includes both uncertainty and negative consequences. This acceptability depends on cultural factors because people's views change according to time and location.

The process of risk assessment in agro-chemical consists of (i) hazard identification, (ii) exposure assessment, (iii) effect's management, (iv) risk characterisation, and (v) risk management. However, transgenic crops may be able to invade (or colonise) and multiply in many habitats. Hence, this risk assessment of a genetically modified living organism (also known as GMLO) must consider other characteristics not included when assessing the release of non-living compounds to the environment, e.g. horizontal gene transfer between transgenic crops and wild related species.

Scientific risk assessment of transgenic crops must be strictly performed and precautionary principles should be considered in the decision making process. In the industrialised world, this precautionary principle is a key component of the response to the unforeseen (and sometimes irreversible) human and environmental impact, which may occur by introducing into the system new advances ensuing from research and technology development. In Norway, an unique legislation advocates that "the production and use of GMO should be ethically and socially justifiable in accordance with the principle of sustainable development" as well as "safe to humans and to the environment". By applying this framework, marketing applications of GMO could be rejected if insufficient documentation regarding ecological and heath aspects was submitted by the producer.

What are the potential ecological risks associated with the release of GMP into the farming system? These are of course a very large number of potential risks, However, perhaps the two most important risks are:

(i) GMP establishes in semi- or natural habitats, and

(ii) inserted transgenes incorporate into other species, thereby affecting non-target organisms in farms or natural habitats.

Hierarchical test protocols have been proposed to assess the risks of releasing GMP. Such protocols require knowledge about evolutionary history, morphology, life-history characteristics, pollination or breeding system, gene-transfer likelihood, natural hybridisation, recruitment and vegetative propagation of a chosen species. Likewise, producers should provide, to facilitate this risk assessment, additional information regarding biochemical, physiological, and morphological changes owing to inserted gene(s), along with a list and description of marker and reporter genes included in the transgenic plant. It would also be important to add details concerning when and in which plant tissues or organs will be expressed the modified function or phenotype.

Nonetheless, people must also know that scientists assessing risks of transgenic crops may extrapolate the outcome or results from simple short-time experiments into complex long-term natural- or farming systems. Investigations about gene flow and competing ability of transgenic crops may be easily addressed through short-term experiments. However, the assessment of the environmental impact of GMP requires a long-term, expensive, holistic research. Computer modelling, which integrates knowledge about gene flow, competing ability, spread of transgenes to weedy species, and cultural practices in the farming system, may provide an alternative means for long-term risk assessment of releasing GMP into the environment.

Consumer concern about transgenic crops also focuses on their safety as food, especially if modifications could influence their metabolism or health. In this regard, transgenic plants without selectable markers, such as antibiotic resistance genes, are needed to convince GMP-sceptics of the advantages of genetic engineering for crop improvement. In this way, their criticism concerning the potential risks of transgenic crops could be overcome. For example, molecular or metabolic markers may provide a means to identify transgenic plants with desired trait(s). Of course, these alternative markers should be safe from an environmental and health perspective.

2

Crop Plants

Transgenic Crops

The products of genetic engineering are often called genetically modified organisms, or GMOs. All these terms refer to methods of recombinant DNA technology by which biologists splice genes from one or more species into the DNA of crop plants to transfer chosen genetic traits. Genes are segments of DNA that contain information that in part determines the structure and function of a living organism. Genetic engineers are molecular biologists who manipulate this information, typically by taking genes from one species "an animal, plant, bacterium, or virus" and inserting them into another species, such as an agricultural crop.

With the advent of genetic engineering of plants around 1983, it appeared that this new biotechnology would benefit and even revolutionise agriculture. The transfer of desirable genetic traits across species barriers has shown promise for solving problems in the management of agricultural crops. Potential benefits include reduced toxic pesticide use, improved weed control resulting in less tillage and soil erosion, and water conservation. However, "emerging evidence suggests the promised environmental benefits remain small, uncertain or unrealised in the U.S., and some risks are real."

Gene Transfer

Knowing how gene transfer occurs in the laboratory is key to understanding how the desired traits express themselves in the resulting crop plants, and why unintended effects, such as the production of allergens, may also be present or develop with subsequent generations of plants. Plants have elaborate defence mechanisms for dealing with foreign substances, including foreign DNA. To overcome these defence mechanisms, genetic engineers include pathological organisms in the complex DNA packages they create to carry the chosen gene into the host cell.

Generally, there are four parts to a DNA package:

1. Genes for the desired trait the "payload". An example of such a trait is crop resistance to a given herbicide.
2. Genes for carrying the package into the host plant's DNA. This genetic carrier is called the vector and is usually taken from a bacterium that causes tumours in plants, Agrobacterium tumefaciens. Viruses are also sometimes used as gene carriers. These bacterial or viral vectors infect the new host cells, delivering the engineered gene into the DNA of the host plant.
3. Genes for ensuring that the genetic package will express the desired trait persistently (rather than weakly or not at all). These genes turn on the desired trait and are called promoters. They are usually derived from the cauliflower mosaic virus (CaMV).

4. Genes for helping the biologist find the DNA segment in which the insertion has been successful. These genes are called markers and are resistant to antibiotics (usually neomycin or kanamycin). When the antibiotic is applied to the new host's cells, the cells that survive are the ones carrying the successfully inserted antibiotic-resistant gene indicating a likelihood that the gene carrying the desired trait has been successfully inserted as well. Marker genes are usually derived from bacteria (E. coli).

An alternative vector method uses a particle gun to shoot tiny metal particles with the DNA package attached into the new host cell. This eliminates the need for a bacterial vector, but adds to the potential for genetic disruption in the host cell.

The current methods of gene transfer are not precise. While scientists can control with relative exactness the "trait gene' that is inserted into a host plant genome, they cannot yet control its location, nor the number of copies that get inserted. Location of genetic material is important because it controls the expression of biological traits just as genes themselves do. Also, inserted DNA frequently contains multiple copies, which are often scrambled. Multiple copies can "silence" genes, resulting in instability in the introduced genes.

Because scientists can control neither the location nor copies of the genetic material they are inserting, they cannot entirely control the traits the host plant will express, nor can they guarantee genetic stability in subsequent generations. This can lead to unpredictable and undesirable effects, examples of which include plant infertility, the production of toxins, and reductions in yield and plant fitness.

Another characteristic that can result in unintended effects is called pleiotropy. Pleiotropy means that one gene controls multiple traits in an organism. In other words, the gene for a desired trait may also control other traits in the crop plant. Genetic engineers have relied on a simpler model of gene expression, assuming one gene controls one particular trait. But pleiotropy is common, and the interactions of genes with each other and their environment add further complexity. In short, it is difficult or impossible to predict what the effects of new genetic combinations will be.

The current marker and promoter genes of choice also create new hazards. The antibiotic-resistant marker genes carry the potential to increase the number of bacteria that are resistant to antibiotics. The viral promoter genes could combine with other infecting viruses to create new viruses. The very powerful promoter commonly used, the cauliflower mosaic virus, can cause the inserted DNA package to be expressed out of proportion with the rest of the genetic code. With the particle gun method especially, this promoter can also jump out of the DNA package and land somewhere else in the host genome, causing disruption. The bacterial and viral vector genes can recombine to form active pathogens once again either new ones, or old ones with renewed virulence, or with broader host specificity.

Each piece of the gene package described above carries with it the potential to disrupt non-target portions of the host plant's DNA, to create instability in the new genetic construct, or to result in unpredictable combinations that can create new substances, viruses, or bacteria. What this adds up to is the possibility of unintended effects, particularly in subsequent generations of the engineered plant.

Genetic engineering of crops is based upon the notion that genes alone determine species traits. But this assumption and the resulting idea of simply transferring a single gene between organisms to achieve a single result contradicts recent discoveries in genetic science. Inheritance and expression of traits are controlled by complex processes involving complicated genetic material that includes more than simple genes.

For example, so-called junk DNA is now thought to play a significant role in determining an organism's traits. Mapping of the human genome, completed in early 2001, revealed that humans contain about one-third the number of genes scientists expected to find. In fact, the 30,000 human genes they identified comprise only 1 percent of the total human DNA; they don't know what much of the other 99 percent is for. Only 300 genes in the human genome are not also in the mouse.

Transgenic Crops and Traits

While increased yields and improved nutritional value are among the promised benefits of transgenic crops, most genetically engineered crops now planted worldwide are designed either to 1) survive exposure to certain herbicides (called herbicide-tolerant crops) or 2) kill certain insect pests (called pesticidal or insecticidal crops). Genetically engineered herbicide-tolerant crops have been altered to withstand being sprayed with broad-spectrum herbicides, with the idea that one application will take care of most types of weeds without killing the crop. Insecticidal crops contain genes of the soil bacterium Bacillus thuringiensis (Bt).

These Bt genes cause the plants to produce a chemical toxic to the European corn borer, the cotton bollworm, and other caterpillars. (Caterpillars are the larvae of insects in the Lepidoptera order, which includes moths and butterflies). Herbicide-tolerant crops accounted for 71 percent of the acreage planted, worldwide, to genetically engineered crops in 1998 and 1999. Pesticidal crops, or a combination of pesticidal and herbicide-tolerant crops, accounted for most of the remaining acreage.

More than 100 million acres of the world's farmland were planted with transgenic crops in the year 2000. The United States, Argentina, and Canada are the world's leading producers of genetically engineered crops, with the U.S. first at 71 percent of the world's total acreage. Transgenic crops are being shipped to or experimented with in many other countries, including China, India, Australia, and South Africa.

In 1999, one-quarter of U.S. farmland was planted to transgenic crops. To date, soybeans and corn cover the most transgenic acres. In 2001, more than 60 percent of soybeans planted in the U. S. were estimated to be transgenic specifically, tolerant to Monsanto's broad-spectrum herbicide, Roundup (glyphosate). Transgenic corn planted in the U.S. offers both insecticidal and herbicide-tolerant traits, though Bt insecticidal corn predominates.

Other large-acreage transgenic crops include cotton and canola. Most transgenic cotton is herbicide-tolerant, though some varieties have the Bt trait; transgenic canola is herbicide-tolerant. The first transgenic wheat, planned for commercial introduction in 2003, is Roundup-tolerant. Other traits engineered into commercial transgenic varieties include disease resistance, high pH tolerance, and several nutritional, taste, texture, and shelf-life characteristics. Other transgenic crops currently on the U.S. market include tomatoes, potatoes, sunflowers, peanuts, and sweet mini-peppers. More transgenic crops, including rice, are under development for commercial release in the next several years.

However, there is evidence that biotechnology firms may be changing course because of deepening consumer skepticism and tighter regulation worldwide, which are increasing the costs and business risks of new transgenic crop introductions. A Monsanto spokesman was quoted in the Christian Science Monitor in August 2001, saying, "We're focusing on four core crops—corn, oilseeds, cotton, and wheat," the major crops in North America with the most acreage and profit potential.

There are many unanswered questions about transgenic crops and their potential benefits, costs, and risks. In fact, according to a recent independent survey of research data on transgenic crops, conducted by the Winrock Foundation's Henry A. Wallace Center for Agricultural and Environmental Policy, .The varieties and uses of genetically altered crops have grown much more rapidly than our ability to understand...them.

Ecological Issues in Transgenic Crops

Gene flow

Ecological scientists have little doubt that gene low from transgenic fields into conventional rops and related wild plants will occur. Gene low from transgenic to conventional crops is of concern to farmers because of its potential to ause herbicide resistance in related conventional crops. For example, in western Canada, three different herbicide-resistant anola varieties have cross-pollinated to create anola plants that are resistant to all three types of herbicide. Through what is called gene-stacking, this new triple resistance has turned volunteer canola into a significant weed problem.

Gene flow from transgenic crops to wild relatives creates a potential for wild plants or weeds to acquire traits that improve their fitness, turning them into "super weeds". For example, if jointed goatgrass—a weedy relative of wheat acquires the herbicide-tolerant trait of Roundup Ready wheat, it will thrive in crop fields unless applications of other herbicides are made. There is already evidence of such outcrossing from herbicide-resistant wheat to jointed goatgrass.

Frank Young and his colleagues at Washington State University found that imidazolone-resistant wheat (not a transgenic variety) outcrossed to goatgrass in one season. Other traits that wild plants could acquire from transgenic plants that would increase their weediness are insect and virus resistance. Because of their experience with classically bred plants, few scientists doubt that genes will move from crops into the wild: seven of the world's thirteen most important crop weeds have been made weedier by genes acquired from classically bred crops. Because gene flow has the potential to affect farmers' crop and pest management, crop marketability, and liability, more research needs to be done to determine the conditions under which gene flow from transgenic plants is likely to be significant.

Pesticide resistance in insect pests Bt has been widely used as a microbial spray because it is toxic only to caterpillars. In fact, it is a pest management tool that organic farmers depend on one of the few insecticides acceptable under organic rules. Unlike the commercial insecticide spray, the Bt engineered into crop plants is reproduced in all, or nearly all, the cells of every plant, not just applied on the plant surface for a temporary toxic effect.

As a result, the possibility that transgenic Bt crops will accelerate insect pests development of resistance to

Bt is a serious concern. Pest resistance to Bt would remove this valuable and environmentally benign tool from farmers and forest managers' pest control toolbox.

Antibiotic resistance

Research is needed on antibiotic resistance management in transgenic crops. Already the European Commission's new rules governing transgenic crops stipulate phasing out antibiotic-resistant marker genes by the end of 2004.

Beneficial organisms

Evidence is increasing that transgenic crops—either directly or through practices linked to their productions are detrimental to beneficial organisms. New studies are finding that Bt crops exude Bt in concentrations high enough to be toxic to some beneficial soil organisms. In addition, a study by University of Arkansas scientists has shown that "root development, nodulation and nitrogen fixation are impaired" in some varieties of Roundup Ready soybeans. The reason is that the beneficial rhizobium responsible for nitrogen fixation in soybeans is sensitive to Roundup. It also appears that disruption of beneficial soil organisms can interfere with plant uptake of phosphorus, an essential plant nutrient.

Beneficial insects that prey on insect pests can be affected by insecticidal crops in two ways. First, the Bt in transgenic insecticidal crops has been shown in some laboratory studies to be toxic to ladybird beetles, lacewings, and monarch butterflies. The extent to which these beneficials are affected in the field is a matter of further study. Second, because the insecticidal properties of Bt crops function even in the absence of an economic threshold of pests, Bt crops potentially can reduce pest populations to the point that predator species are negatively affected.

Reduction of genetic diversity

As fewer and larger firms dominate the rapidly merging seed and biotechnology market, transgenic crops may continue the trend toward simplification of cropping systems by reducing the number and type of crops planted. In addition, seed-saving, which promotes genetic diversity, is restricted for transgenic crops.

Food Safety

Because food safety is predominantly a concern of consumers, this paper does not include a discussion of it. However, because food safety issues do impact marketing and international trade, here is a short list of consumer safety concerns about transgenic food:

- possibility of toxins in food
- possibility of new pathogens
- reduced nutritional value
- introduction of human allergens
- transfer of antibiotic resistance to humans' unexpected immune-system and genetic effects from the introduction of novel compounds. It is in part because of these concerns that consumer demand for organically grown crops continues to increase.

Pest Management for Farmers

The most widely planted transgenic crops on the market today can simplify short-term pest management for farmers and ranchers. In the case of herbicide-tolerant crops, ideally farmers can use a single broad-spectrum herbicide for all their crop weeds, though they may need more than one application in a season. By planting insecticidal crops, farmers can eliminate the need to apply pesticides for caterpillar pests like the European corn borer or the cotton bollworm, though they still have to contend with other crop pests.

While these crops offer handy pest control features, they may complicate other areas of farm management. Farmers who are growing both transgenic and conventional varieties of the same crop will need to segregate the two during all production, harvesting, storage, and transportation phases if they are selling into differentiated markets or plan to save their own seed from their conventional crops.

To minimise the risk of gene flow from transgenic to adjacent conventional crop fields, federal regulations require buffer strips of conventional varieties around transgenic fields. Different transgenic crops require different buffer widths. Because the buffer strips must be managed conventionally, producers have to be willing to maintain two different farming systems on their transgenic fields. Crops harvested from the buffer strips must be handled and marketed as though they are transgenic.

Planted refuges where pest species can live outside fields of insecticidal and herbicide-tolerant transgenic crop' are also required to slow the development of weed and insect pest resistance to Bt and broad-spectrum herbicides. These refuges allow some individuals in the pest population to survive and carry on the traits of pesticide susceptibility. Requirements governing the size of refuges differ according to the type of transgenic crop grown. Farmers growing herbicide-tolerant crops need to be aware that volunteer crop plants the following year will be herbicide-resistant.

This herbicide resistance makes no-till or direct-seed systems difficult because volunteers can't be controlled with the same herbicide used on the rest of the crop. In a no-till system that relies on the same broad-spectrum herbicide that the volunteer plants are resistant to, these plants will contaminate the harvest of a following conventional variety of the same crop, a situation farmers need to avoid for two reasons. First, the contamination means a following conventional crop will have to be sold on the transgenic market. This leads to the second reason.

If farmers grow and market a transgenic crop for which they do not have a technology agreement and did not pay royalty fees, they face possible prosecution by the company that owns the transgenic variety. Many farmers have been charged with "theft" of a company's patented seed as a result of contamination in the field. Farmers growing insecticidal crops need to recognise that insect pressure is difficult to predict and may not warrant the planting of an insecticidal variety every year. In a year when pest pressure is low, the transgenic seed becomes expensive insurance against the threat of insect damage.

Farmers growing transgenic crops need to communicate with their neighbors to avoid contaminating their fields and to ensure that buffers are adequate. In Maine, farmers growing transgenic crops are now required by law to be listed with the state agriculture department, to help identify possible sources of cross-contamination when it occurs. The new law, L.D. 1266, also requires manufacturers or seed dealers of genetically engineered plants, plant parts or seeds to provide written instructions to all growers on how to plant, grow, and harvest the crops to minimise potential cross-contamination of non-genetically engineered crops or wild plant populations.

Farm management issues common to all transgenic crops include yield, cost, price, profitability, management flexibility, sustainability, market acceptance, and liability.

Crop Yield, Costs, and Profitability

Some farmers will get higher yields with a particular transgenic crop variety than with their conventional varieties, and some will get lower yields. The same is true for costs. Some producers will see their overall profitability rise, and some will see it drop. If the results of research studies so far on yield, chemical use, costs, crop prices, and profitability of these crops seem contradictory, it is because farmers and farms are not all alike. Neither are the conventional varieties used for comparison.

Some yield, cost, and profitability trends do appear to be emerging from the growing body of research data for transgenic crops, however. Herbicide-tolerant soybeans appear to suffer what's referred to as a yield drag. Again, in some areas and on some farms this tendency of Roundup Ready soybean varieties to yield less than their comparable, conventional counterparts varies, but overall, they appear to average yields that are five to ten percent lower per acre. Drought conditions worsen the effects. The bacterium that facilitates nodulation and nitrogen fixation in the root zone apparently is sensitive to both Roundup and drought.

Yields of herbicide-tolerant cotton are reportedly not significantly different from those of conventional cotton. Herbicide-resistant transgenic canola varieties yield less on average than conventional canola varieties. Transgenic canola costs less than conventional canola to produce, but because of its higher yields conventional canola returns more profit per acre.

Insecticidal Crops—Corn, Cotton

Insecticidal Bt corn and cotton generally yield higher in most years for some regions' according to USDA

Economic Research Service data from 1996 to 1998. Bt cotton, especially, outpaces yields of conventional cotton by as much as 9 to 26 percent in some cases, though not at all in others. Yield increases for Bt corn have not been as dramatic. Time will tell whether farmers can expect yield increases or decreases in the long run with these and other transgenic crop varieties.

Use of Chemical Pesticide

One of the promises of biotechnology is that it will reduce pesticide use and thereby provide environmental benefits and reduce farmers' costs. The herbicide-tolerant and insecticidal varieties are designed specifically to meet these goals. Studies estimate a two to three percent decrease in U.S. pesticide use, but the effects vary widely by crop, region, and year. Increased future pesticide use resulting from the buildup of resistance to heavily used herbicides is a long-term concern. Pesticide use depends on the crop and its specific traits; weather; severity of pest infestations; farm management; geographic location of the farm; and other variables. As a result, conclusions drawn by various studies analysing pesticide use on transgenic crops remain controversial.

Herbicide-tolerant cotton has also resulted in little change in herbicide use. The data for herbicide-tolerant soybeans seems harder to sort out. A recent study of herbicide use data on Roundup Ready soybeans by Dr. Charles Benbrook, former executive director of the National Academy of Sciences Committee on Agriculture and now with the Northwest Science and Environmental Policy Center, concludes that the use of herbicides has actually increased because the weeds have become resistant to Roundup.

While another recent study by Netherlands scientists shows a decrease in herbicide use on transgenic soybeans, it is clear that weed resistance to Roundup may lead to increased herbicide use and to the need to shift to more toxic compounds in the future. American Soybean Association president Tony Anderson agrees that the developing resistance of weeds to herbicides such as Roundup is a problem.

The Wallace Center report emphasises the importance of ongoing monitoring of pesticide use data. If farmers abandon integrated pest management, which utilises a variety of pesticide and cultural control methods, in favour of the simplified control offered by herbicide-resistant and insecticidal transgenic crops, then early findings of reduced pesticide quantities and toxicity may not hold over the long run.

Profitability

Farmers need to consider all the factors that determine profitability. No single factor can tell the whole story. Transgenic crop seeds tend to be more costly, and farmers have the added expense of a substantial per-acre fee charged by the owners of transgenic varieties. These costs have to be considered along with input cost changes—whether herbicide or insecticide use and costs go down, go up, or stay the same. Market price is another factor: Prices for some transgenic crops in some markets are lower than prices for comparable conventional crops, though rarely are they higher.

Farmers need to watch the markets. Some buyers will pay a premium for a non-transgenic product, though as transgenic seeds find their way into conventional transportation, storage, and processing streams, these premiums may disappear along with confidence that "GMO-free" products are in fact truly free of engineered genes.

Marketing and Trade

Buyer acceptance is a significant marketing issue for farmers raising transgenic crops. Farmers need to know before they plant what their particular markets will or won't accept. Since most grain handlers cannot effectively segregate transgenic from non-transgenic crops in the same facility, many companies are channelling transgenic crops into particular warehouses. Farmers need to know which ones and how far away those are.

Many foreign markets tend to be more leery of transgenic products than domestic markets, though trade in transgenic livestock feed is more liberal than trade in transgenic human food. The widespread contamination of corn in the U.S. with the Cry9C Bt transgene (StarLink), which is not approved for human consumption, has resulted in even further resistance on the part of buyers to purchasing transgenic products for human food. According to a report in Britain's *The Guardian*—No new transgenic crops have been approved by the European Union (EU) since April 1998, and a defacto moratorium on further approvals has been in place since June 1999.

Some countries have banned altogether the production and importation of transgenic crops and food products (Sri Lanka and Brazil), while many others have put in place partial bans or mechanisms to slow their approval for importation or production. While the European Union officially lifted its two-year moratorium on the introduction of new transgenic crops in February 2001, during the debate over labelling and traceability regulations the moratorium remains in effect. Under the proposed new EU requirements, "all foods and animal feed derived from GMOs have to be labelled and, in the case of processed goods, records have to be kept throughout the production chain allowing the GMO to be traced back to the farm". If approved, the new regulations will hamper the export of U.S. farmers' products to the EU because the U.S. does not require traceability or labelling of transgenic crops.

While the U.S. does not require mandatory labelling of processed food containing transgenic ingredients, the European Union, Russia, Czech Republic, Japan, South Korea, Taiwan, Australia, New Zealand, and Ecuador do. Because many domestic merchandisers of agricultural commodities do not segregate transgenic from conventional crop varieties, it is impossible for them and the farmers that supply them to serve these food markets. Twenty percent of corn and 8 percent of soybeans produced in the U.S. are exported, and more than 80 percent of these crops is used in animal feed. Overseas market restrictions on transgenic crops in the feed sector are less stringent than those for food.

In contrast, 48 percent of U.S. wheat is exported, with nearly half of this food crop going to countries that are resistant to buying transgenic food, particularly Japan and European nations. Because wheat producers are so dependent on exports, they are approaching the introduction of the first transgenic wheat, expected as soon as 2003, with caution. The Japanese milling industry has made it clear that it does not want transgenic products. As a result, Monsanto has promised not to introduce its Roundup Ready wheat until Japan gives its approval.

In addition to national and international policies on the use and importation of transgenic crops, many countries. processors and retailers have set their own corporate policies. Major retail chains in Europe and the U.S. have declared their commitment to avoiding the purchase of transgenic products, both feed and food. For buyers who don't want transgenic products, the likelihood of contamination of non-transgenic products with engineered genes is troublesome.

Since 1997, the European Union has effectively barred U.S. corn imports over the possibility that genetically engineered varieties unapproved in the EU have mixed with sanctioned crops. This has cost American farmers access to a $200 million-a-year market. Organic farmers face even bigger marketing and trade risks, since their buyers can accept no transgenic contamination. The organic industry does have a system for segregation, but recent tests for transgenic material in organic products demonstrate that it is not immune to contamination from conventional systems.

The markets organic farmers have enjoyed, and those that producers of non-GE conventional crops could build upon, will likely disappear if contamination of their products continues to spread. Until the issues of trade in transgenic crops and food products, and their labelling and traceability, are resolved in international trade arenas, the markets for U.S. transgenic crops will remain uncertain. Changes in what any particular country will or won't accept are occurring all the time, and producers of crops for export are well advised to stay informed of these changes. Segregation of transgenic crops and liability for cross-contamination of conventional and organic crops are issues that obviously overlap with those of marketing and trade.

Liability

The need to separate transgenic crops from conventional and organic ones opens farmers to liability for their product at every step from seed to table. Effective systems for segregation do not exist at present, and will be costly to develop and put into place. Farmers may well end up bearing the added costs of crop segregation, traceability, and labelling. In the meantime, farmers who grow transgenic varieties and, ironically, those who do not are liable for transgenic seeds ending up where they aren't wanted: in their own non-transgenic crop fields, in neighbours' fields, in truckloads of grain arriving at the elevator, in processed food products on retail shelves, and in ships headed overseas.

Farmers who choose not to grow transgenic varieties risk finding transgenic plants in their fields anyway, as a result of cross-pollination via wind, insects, and birds, which may bring pollen from transgenic crops planted miles away. Besides pollen, sources of contamination include contaminated seed and seed brought in by passing trucks or wildlife. Those farmers whose conventional or organic crops are contaminated, regardless of the route, risk lawsuits filed against them by the companies that own the proprietary rights to seed the farmer didn't buy. Likewise, farmers who grow transgenic crops risk being sued by neighbours and buyers whose non-transgenic crops become contaminated.

Because contamination by transgenic material has become so prevalent in such a short time, all farmers in areas of transgenic crop production are at risk. Insurance, the most common recourse for minimising potential losses because of liability, is not available to the nation's farmers for this risk because insurance companies do not have enough information to gauge the potential losses. Most of the farmers who have been accused by transgenic seed companies of illegally growing and harvesting their proprietary transgenic varieties have paid fines to the companies rather than go to court to defend themselves.

One Canadian grower of non-transgenic canola, Percy Schmeiser, did go to court against Monsanto, and lost in a recent ruling that requires him to pay Monsanto the approximately US$85,000 value of his crop and $13,000 in punitive damages. In another example, a Texas organic farm's corn, assumed to be GE-free, was purchased by a processor who made it into organic tortilla chips. Only after the product had been sold and shipped to European retailers was it discovered to be contaminated with transgenic corn. The processor had to recall its product at a cost of over $150,000. The processor chose not to sue the organic farmer, but could have. The retailer, in turn, apparently did not sue the processor. Had it done so, the liability for the retailer's loss could have fallen on both the processor and the farmer.

According to a Boston Globe story, no figures on farmers' losses resulting from liability for transgenic contamination exist, but anecdotal evidence suggests that cases against farmers are becoming far more common.. Until laws or legal precedent clarify the extent of farmer liability, farmers would do well to avoid making assumptions or claims about the purity of their non-transgenic products. Furthermore, producers of transgenic crops need to take all possible precautions against spreading pollen and seed to their own and others' non-transgenic fields and markets.

A full risk assessment and legal clarification of the distribution of liability among farmers, seed companies, grain handlers, processors, and retailers is needed before farmers can rest assured that transgenic crops won't result in lawsuits against them. For farmers who have adopted transgenic technology and for those who have not in the words of Brian Leahy, executive director of California Certified Organic Farmers, .This technology does not respect property rights. In essence, farmers who grow transgenic crops on some of their fields, and farmers who grow none, risk bearing tremendous liability. This situation won't change until farmers either gain legal protection or until they all grow transgenic crops on all their fields.

Research on Transgenic Crops

While transgenic crop varieties are generally the property of private corporations, those corporations often contract with public-sector agricultural research institutions for some of their development work. In fact, private investment in agricultural research, including germplasm development, has surpassed public investment in recent years. With this shift in funding sources, the following questions become important: Is the private sector unduly influencing the public research agenda? Are corporations directing public research in less socially valuable directions while research on, for instance, sustainable agriculture goes wanting? Are the outcomes of corporate-funded transgenic research and development by our public institutions equitable across the food and agricultural sectors? Is equity even a consideration of our public institutions when they accept this work?

Because intellectual property rights (patents) apply to living organisms, making them private property, the free flow of scientific information that has historically characterised public agricultural research is being inhibited. What are the implications for the future of agriculture and society of the secrecy that now surrounds so much of what was formerly shared public knowledge? These and other questions need to be addressed by citizens and their public institutions. These issues are of particular concern to farmers and consumers who would benefit from research into .alternative technologies that are less costly (in every way), less risky, and more equitable. Equity requires that the economic benefits and risks of transgenic technology be fairly distributed among technology providers, farmers, merchants, and consumers.

For long-term sustainability, farmers need research that focuses on farms as systems, with internal elements whose relationships can be adjusted to achieve farm management goals. In contrast, transgenic crop research so far has focused on products that complement toxic chemical approaches to control of individual pest species. These are products that can be commercialised by large agribusiness or agri-chemical interests and that farmers must purchase every year. This research orientation only perpetuates the cost-price squeese that continues to drive so many out of farming.

Seed and Biotech Industries

The broadening of intellectual property rights in 1980 to cover living organisms, including genes, has resulted in a flurry of mergers and acquisitions in the seed and biotech industries. According to the Wallace Center report, .Relatively few firms control the vast majority of commercial transgenic crop technologies. These firms have strategically developed linkages among the biotechnology, seed, and agri-chemical sectors to capture as much market value as possible. However, these tightly controlled linkages of product sectors raise serious issues of market access, product innovation, and the flow of public benefits from transgenic crops.....

Unlike Plant Variety Protection which does not allow for the patenting of individual genes, but only of crop varieties—Intellectual Property Rights prohibit farmers from saving their seed and undertaking their own breeding program, and prohibit plant breeders from using the material to create a new generation of varieties adapted to specific regions or growing conditions.

By 2000, agri-chemical giants DuPont and Monsanto together owned 73 percent of the corn seed producers in the U.S. This kind of corporate control and concentration raises the question of whether there remains enough competition in the seed industry for seed pricing to remain competitive. As additional concentration occurs, how affordable will seed all seed, not just transgenic seed be for farmers? This question takes on added gravity as an increasing number of seed varieties become proprietary. Farmers can't save proprietary seed for planting and so must purchase new seed every year. In addition, farmers choosing transgenic varieties must sign a contract with the owner of the variety and pay a substantial per-acre technology fee, or royalty.

The anticipated commercial introduction of transgenic wheat represents a dramatic shift in an industry in which farmers still widely save their own seed. As non-transgenic varieties become contaminated with transgenic ones, even those farmers who choose to stick with conventional varieties will lose the right to replant their own seed. This loss has already occurred in Canada's canola industry, with Monsanto winning its court case against farmer Percy Schmeiser for replanting his own canola variety that had become contaminated with Monsanto's Roundup Ready canola.

The adoption of transgenic crop varieties has brought with it an increasing prevalence of contract production. While contract production can lead to increased value and reduced risk for growers, farmers are justified in their concern about their loss of control when they sign a contract with a private company. Issues associated with contract production of transgenic crops must be considered within the broader context of a sustainable agriculture to include ownership, control, and social equity.

Regulation of Transgenic Crops

Much of the controversy over transgenic crops, both internationally and in the U.S., is in part a result of how the U.S. regulates transgenic crops. The federal government has determined that the commercial products of agricultural biotechnology are 'substantially equivalent' to their conventional counterparts and that therefore no new regulatory process or structure is needed for their review and approval. Currently, three federal agencies regulate the release of transgenic food crops in the U.S.: the U.S. Department of Agriculture's Animal and Plant Health Inspection Service (USDA-APHIS), the U.S. Environmental Protection Agency (EPA), and the U.S. Food and Drug Administration (FDA).

USDA-APHIS: The USDA looks at how a transgenic plant behaves in comparison with its unmodified counterpart. Is it as safe to grow? The data it uses are supplied largely by the companies seeking a permit for release of the new crop. Under 'fast-track' approval, a process in place since 1997, companies introducing a crop similar to a previously approved version need give only 30 days advance notice prior to releasing it on the market.

EPA: The EPA regulates the pesticides produced by transgenic crops, such as the Bt in Bt corn and cotton. It does not regulate the transgenic crops themselves. In contrast to its regulation of conventional pesticides, the

EPA has set no tolerance limits for the amount of Bt that transgenic corn, cotton, and potatoes may contain.

FDA: The FDA focuses on the human health risks of transgenic crops. However, its rules do not require mandatory pre-market safety testing or mandatory labelling of transgenic foods.

Initially, the U.S. regulatory process for transgenic food crops required product-by-product reviews. Now, however, to simplify and speed up the process, new products can be approved based on the experience gained in reviewing earlier products. According to the Wallace Center report, the implication is that .some crops might be approved, or disapproved, without actual field testing. The regulatory process, in fact, may not answer most questions about the environmental and human health risks of commercial production of these crops.

Central to the policy of substantial equivalence is the assumption that only the end product of transgenic technology is of concern not the process of genetic modification. Canada has adopted a similar approach. Europe and other U.S. trading partners, however, have taken a more conservative approach. They focus on the process of genetic modification, the source of many of the environmental and human health risks of greatest concern.

How these different approaches play out in reality can be summed up simply: The U.S. and Canada assume a product is safe until it is proven to carry significant risk; the European Union, which follows the precautionary principle, assumes the same product may carry significant risk until it can be proven safe. The science used by the two approaches is not fundamentally different. The difference is in the level of risk the different societies and political systems are willing to accept.

Transgenic Crops and Sustainable Agriculture Development

In contrast to the ecological approach of sustainable agriculture, the current generation of transgenic crops follows a pest management model like that employed for chemical pesticides "through interventions that are toxic to pests,' the Wallace Center report points out. This single-tool approach is likely to fail in the long run because pests will successfully develop resistance that allows them to thrive.

Standard plant-breeding methods can potentially solve many of the same problems in agriculture that genetic engineers are working on, though there are areas in which genetic engineering can enhance traditional plant breeding. Armed with the map of an organism's genetic code, scientists can test which genes are in a plant to select more easily which ones to cross-breed.

The sustainability of any agricultural technology can be gauged in part by answering a series of questions that emerge from the principles of sustainable agriculture. Farmers and ranchers can ask themselves these questions in the context of their own operations to help determine whether adoption of the technology will move them away from, or toward, increased sustainability.

1. Does the technology increase genetic diversity?
2. Does it maintain a positive balance of pests and predators?
3. Does it protect or enhance soil biota?
4. Does it decrease the quantity or concentration of toxins released into the environment?
5. Does it decrease soil erosion?
6. Does it protect non-target organisms?
7. Does it help protect natural habitats?

8. Does it reduce pest populations and viability?
9. Does it increase farmers' yields? Decrease farmers' costs?
10. Does it increase farmers' market control? Management flexibility? Time?
11. Does it provide benefits to consumers? Will consumers accept it?
12. Does it help citizens globally gain better access to food?
13. Does it protect the public's access to information and improve public trust in agriculture?

3

Fiber Crops

Fiber crops are field crops grown for their fibers, which are used to make paper, cloth, or rope. These crops are generally harvestable after a single growing season, as opposed to trees which are typically grown for many years before being harvested for wood pulp fiber. In specific circumstances, fiber crops can be superior to wood pulp fiber in terms of technical performance, environmental impact or cost. Botanically, the fibers harvested from many of these plants are *bast* fibers; the fibers come from the phloem tissue of the plant. The other fiber crop fibers are seed padding, leaf fiber, or other parts of the plant.

New Fiber Crops

The effects of fiber morphology on paper properties are applicable to any type of fiber, regardless of source. The forest crop is renewable is important, but more important, each crop can be better than the preceding one. It has been only in the last 90 years or so that softwoods and hardwoods have been used for pulp and paper. These fiber types now comprise 94% of the market. Before this, fibers used for pulp and paper products came from cotton, flax, wool, straw, and other fibrous sources (Table 1).

Table 1. Fiber dimensions of various plant materials.

	Fiber dimensions	
	Length (mm)	Diameter (um)
Plant material		
Bamboo	2.7	15.0
Cotton	18.0	20.0
Cotton linters	7.0	20.0
Bagasse	1.7	18.0
Esparto	1.1	9.0
Kenaf	5.5	20.0
Jute	2.0	20.0
Hemp (Manila)	5.9	22.0
Papyrus	1.0	—
Sisal	3.2	20.0
Straw		
rice	1.5	11.0
wheat	1.6	13.0
maize	0.75	—
Softwoods	3.6	35.0
Hardwoods	1.2	25.0

Many areas of the world do not possess indigenous supplies of wood fiber. In these areas-North Africa, Central Asia and the Middle East-demand for paper will grow, and it is important to find supplies of papermaking fibers other than wood. Plenty of non-wood plant fiber is available in these areas. For example, China is the biggest producer of straw pulp in the world. Over 60% of the 5 x 106 tons of paper per year produced in their 1000 mills (Table 2) comes from straw, mostly wheat.

Regardless of fiber source, the effect of morphology will be the same. The subject of fiber-to-fiber bonding has always been recognised for its importance in the area of sheet structure and its effect on paper properties. There are two types of fiber bonding associated to a fibrous network structure: (1) intra-fiber bonding, mainly for fiber stiffness, and (2) inter-fiber bonding, which is mainly for sheet strength. Probably the most influential aspect of how a sheet bonds together is the effect of fiber morphology We know that the properties of any given paper product made of fiber from wood *A* will often vary markedly from an identical product made of fiber from wood *B*, even though the conversion variables are identical.

Table 2. Mill number, production, and per capita consumption in various parts of world (1986-87 data) (Pulp and Paper International 1986-87).

	Number of		Annual production (thousands of metric tons)		
Location board mills	Paper and mills	Pulp board	Paper	Pulp	Per capita consumption (kg)
North America	637	257	79,650	73,624	283
United States	561	223	64,416	51,938	290
Canada	76	34	15,234	21,686	211
Europe	1,588	451	67,761	42,480	84
Asia	2,862	142	41,084	18,871	16
Latin America	425	109	9,942	6,808	26
Australia	24	24	2,313	1,980	121
Africa	90	29	2,382	2,369	6
Total World	5,626	1,012	203,132	146,132	42

These differences in properties can be attributed to the morphological difference between the two wood species. Understanding the influence that pulp fiber morphology has on paper properties has great potential value to the pulp and paper industry Especially so, in projecting their effect on products or, evaluating new crop sources for their fiber supply. Instead of costly and time-consuming pulping evaluations, it may be possible to use selected fiber measurements to estimate the papermaking potential from unutilised tree components, to give direction to tree improvement work by clarifying the fiber properties most selected for genetic control and, to possibly predict the most efficient blend of pulps to achieve a desired set of product properties.

Characteristics of Fiber Crops

To illustrate how fiber morphology can affect paper properties, a few fiber characteristics have been selected as follows.

Fibril angle

It has been shown that fibril angle is an important factor in determining the strength properties of individual fibers.Ffibril angle does not influence sheet strength properties. There is, however, a strong correlation between fibril angle and the stretch properties of pulp sheets (Fig. 1). The data in this figure of unbeaten softwood kraft fiber shows that 79% of the variation in stretch is associated with fibril angle. For a sheet of unbeaten fibers, the stretch is controlled almost exclusively by the extensibility of fiber segments between fiber-to-fiber bonded areas. That is, in the unbeaten sheet, the bonded areas are fewer and the bonds weaker than those in a sheet made from beaten fibers.

The bonds are, however, sufficient to hold the fibers while the fibers themselves stretch by an amount dependent upon fibril angle to a point at which the load can no longer be maintained by the bonded area. In a sheet made from beaten fibers, a greater number of bonded areas are available and the bonds formed are stronger than in a sheet made from unbeaten fibers. Therefore, after beating, the dependence upon fibril angle is lessened (40%), but fibril angle remains the dominant variable. Multiple regression analysis shows that 75% of the variation in stretch can be accounted for by fibril angle and by the numbers of fibers per gram of pulp.

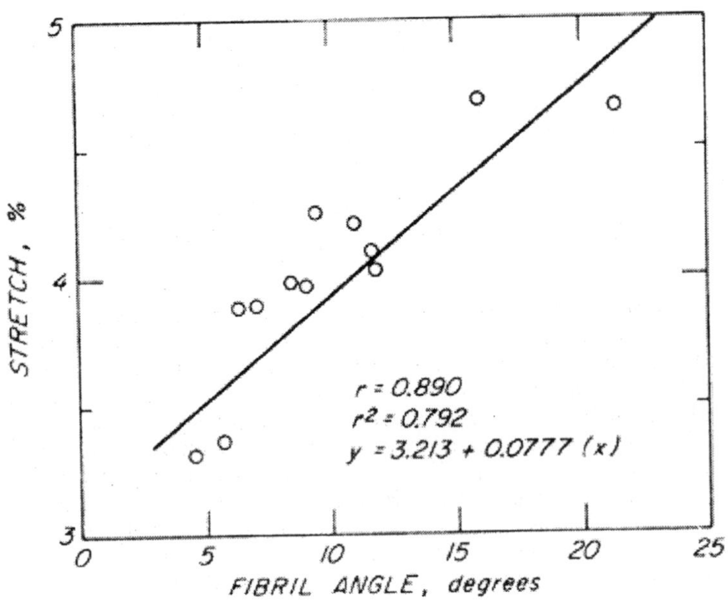

Figure 1. Influence of fibril angle on the stretch of pulp sheets made of unbeaten, unbleached softwood kraft pulp fibers

In the case of unbeaten hardwood pulps, 78% of the variation in stretch properties can be accounted for by fiber length and 70% by fibril angle. However, after beating, the effect of fiber length is less and fibril angle becomes the dominant single variable. Although the dependence on fibril angle is lessened, it accounts for 45% of the variation in stretch of sheets made from beaten hardwood fiber. Multiple regression showed that 92% of the variation in stretch of sheets made from unbeaten pulps could be accounted for by fiber length and fibril angle. The same two fiber characteristics accounted for 75% of the variation in unbeaten pulps.

Fiber length

With softwood fibers, fiber length *per se* does not show a relationship to any one sheet strength property. In the case of hardwood pulps, fiber length can have a significant effect on many sheet properties.

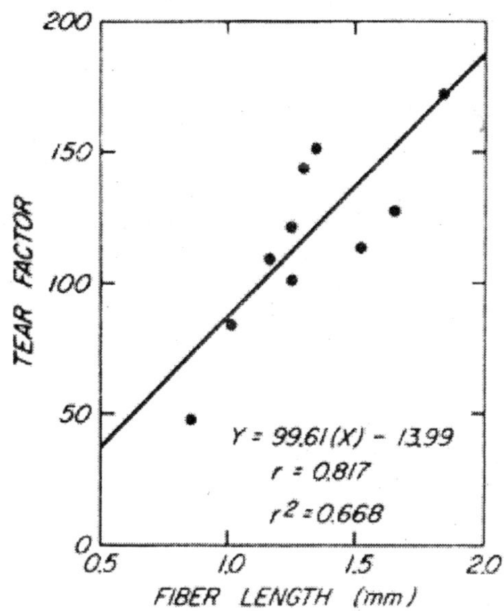

Figure 2. Influence of fiber length on the tearing resistance of pulp sheets made of unbeaten, unbleached hardwood kraft pulp fibers.

A case in point is the effect of fiber length on the tearing strength of pulp sheets. The tearing strength of sheets made from either unbeaten (67%) or beaten (69%) hardwood fiber is principally dependent upon fiber length (Fig. 2, 3).

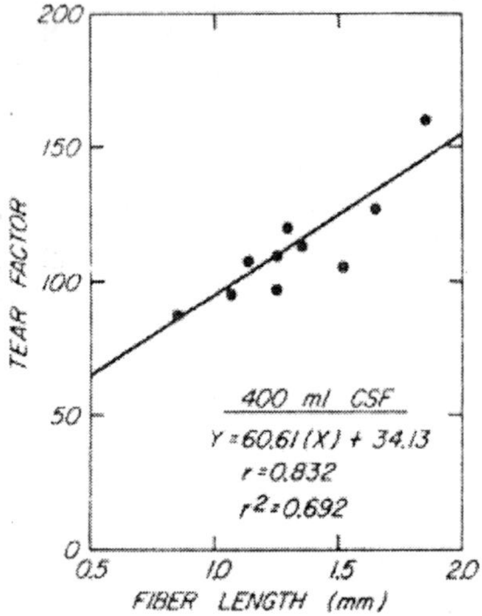

Figure 3. Influence of fiber length on tearing resistance of pulp sheets made of unbleached hardwood kraft pulp fibers (beaten to 400 ml CSF).

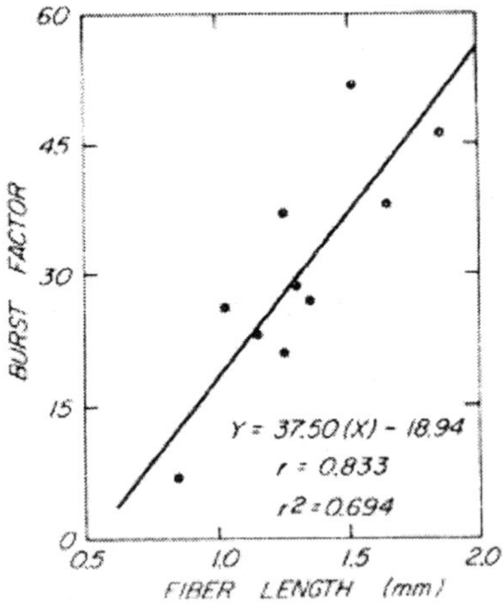

Figure 4. Relationship of bursting strength to fiber length of pulp sheets

The effect of fiber length respect to stretch properties. Bursting strength of sheets made from unbeaten hardwood pulps also show a dependence upon fiber length (Fig. 4).

L/T Ratio

Strength properties of paper, such as bursting strength and tensile strength, are highly dependent upon fiber-to-fiber bonding. Generally, bursting and tensile strengths of paper made from softwoods respond to the same fiber characteristics as do hardwoods. This is especially true after the pulps have been beaten. Visually, it is easy to grasp the significance of cell wall thickness on paper properties, in particular, to those in which bonding is important.

The L/T ratio accounts for 92% of the variation in tensile strength of unbeaten softwood pulps and 84% of the variation after beating. Using L/T as a measure of fiber flexibility is an excellent indicator for the potential use of a softwood pulp for paper products requiring high strength. The bursting strength of unbeaten hardwood pulps is shown to be influenced by fiber length. The dependence of unbeaten hardwood pulp on fiber length may be influenced by the heterogeneity of these pulps. Ifs been shown that the presence of a high percentage of parenchyma cells can reduce sheet strength by inhibiting fiber-to-fiber bond formation. Therefore, the parenchyma cell content of any new fiber source should be a consideration in assessing its potential use in papermaking.

However, after beating, the L/T ratio becomes the dominant factor for both bursting strength and tensile strength of sheets made from hardwood pulps. This most probably reflects the greater degree of fiber collapse which results from beating. The fibers become more flexible and conformable, which in turn provides for more area to be developed for bonding along the fibers length. Also, beating reduces the identity of parenchyma cells and, in a sense, adds bonding material to the furnish. Therefore, bursting and tensile strength, being dependent upon the formation of fiber-to-fiber bonds, is greatly influenced by fiber length and cell wall thickness.

Parenchyma cells and vessel elements

The anatomical makeup of hardwood pulps is much more complex than for softwood pulps. Generally, most hardwoods have much greater amounts of short parenchyma cells than do softwoods and, they have a component not present in softwoods called vessel elements.

Table 3. Effect of vessel elements on tensile strength

			Tensile strength (psi)	
Species[z]	Vessel elements[y] (%)	Unbeaten	Beaten 400 ml CSF	Beaten 400 ml ÇSF
White oak	0	4,150	9,350	—
	1.9	4,200	9,500	9,400[x]
Red alder	0	8,500	16,300	—
	3.7	8,350	16,650	15,900[x]

[z]Fiber fraction used contained 0 percent fines (parenchyma cells).
[y]By weight of original pulp.
[x]Fiber fraction beaten separately and unbeaten vessels added to furnish

Parenchyma cells do affect the strength of pulp sheets. There is only a slight improvement in strength with the removal of parenchyma cells from a white oak kraft pulp, This can, however, be attributed to the poor bonding of thick-walled short fibers characteristic to oak. If, on the other hand, those same parenchyma cells removed from the oak are added to a low parenchyma content red alder furnish, the result is the lowering of the bursting and tensile strengths. This reduction in strength occurs even though the cell wall thickness of red alder is considerably less than that of the oak.

The percentage by weight in hardwood pulps is extremely low and, not surprisingly, their effect on sheet strength is minimal. Table 3 shows the effect of vessel elements on tensile strength. The presence or absence of vessel elements at the percentages found in the original pulp furnish has little influence on the ultimate tensile strength of hardwood pulps.

Woody Fiber Crops

Unlike modern agriculture, forestry in the United States is not dependent on exotic germplasm. Some exotic woody germplasm is planted in forestry but such germplasm represents only a small percentage of the current planting program. In terms of modern forestry the United States is rich in woody plants with more than 2,000 different native species having potential use. Currently, less than a few hundred of these native species are actually utilised for their fiber while a great many more are harvested as a fuel source, for protection plantings, and for use as ornamentals. In the last 25 years, an array of these native woody plants has been investigated to a limited degree to determine their genetic variation.

With the increasing demand for more traditional wood products, including paper and fuelwood, there is a need and considerable opportunity in some parts of the United States to grow and utilise fast-growing, short-rotation woody trees. Various species and hybrids of poplars (*Populus* spp.) and several other native hardwoods have played this role. However, poplars are highly specific as to their site and growing requirements which restrict their potential use to selected areas. The same is true for the other hardwoods, such as American sycamore (*Platanus occidentalis* L.) and green ash (*Fraxinus pennsylvanica* Marsh.).

Since the late 1960s, a number of additional woody species has been selected and tested under accelerated growing conditions. Eucalyptus (*Eucalyptus* L.'Herit) and Royal paulownia (*Paulownia* Sieb. and Zucc.) are exotics, having been introduced in the United States during the 1800s. Both are fast growing hardwoods with many useful qualities. The full potential of these genera are not fully understood nor appreciated, yet there is considerable information that suggests that with a focused effort, both genera would make a useful contribution.

Eucalypts are restricted in range because all members of this genus are susceptible to low temperatures. Yet, with careful selection of species and proper seed source, there are genotypes that can be of considerable value and use. *Paulownia* has had relatively poor acceptance because it is considered difficult to grow and often has poor form. There is virtually no industrial experience with this genus in the United States because of relatively low volumes now available. Yet in Asia, especially in Japan and the People's Republic of China, it is widely used in fine furniture. It is widely grown in its native land of China for its timber value, and as a windbreak tree.

Eucalypts Spp.

Members of this genus are among the most planted forest trees in the world. The genus comprises approximately

530 species and 140 varieties with new species and varieties still being described. It was introduced initially into California and the Hawaiian Islands about 1853. Eucalypts are today planted as ornamentals, as part of shelterbelts, and for fiber and fuelwood in California, Hawaii Arizona, Florida, and New Mexico. They have been planted in a number of southern states with very little success. Since members of this genus have been grown in the United States for over 100 years, why should it be considered a new crop? It is only in the last 25 years that serious studies have been undertaken to understand and identify the extent of the generic variation within this genus. The potential of this genus has not been fully exploited in the United States.

Early introductions were essentially random. Although a number of plantings took place in California during the late 1800s, most of the species and seed sources were eventually found to be unsuitable for timber products. Still, the various species found a use as a windbreak tree or as an ornamental. By the 1960s, more than 150 species were being grown in California alone as ornamentals because of their decorative flowers and pleasing form and shape.

A Eucalypts species introduction and testing center was established in south Florida in 1899 by what is now the USDA Forest Service. The goal of this center was to identify an array of new, fast-growing hardwood species. A few years later, a similar center was established in southern California. Both centers lasted approximately 10 years; but sudden low winter temperatures in both California and the southern United States abruptly ended the wide use of *Eucalyptus* at that rime. Since then, a large series of new species introductions have been carried out, usually with limited success. The collections were often poorly identified and seed source sampling was inadequate.

Investigations of the last 25 years have clearly demonstrated considerable genetic variation within this genus. Although no truly cold-hardy species have been identified, it is still possible to select for a degree of low temperature tolerance in certain species and seed sources. In field studies in California, a number of high-elevation seed sources from Australia withstood, in 1973, a cold wave occurring in late December and January when the temperatures went below freezing on 10 consecutive nights with a low of -7°C. In Florida, selected seed sources have tolerated temperatures as low as -12°C.

With the proper genetic and management studies, it should be possible and feasible to match species and seed source to the appropriate growing conditions. In recent years, this has led to reports of fiber yields of 15 to 25 dry tons per hectare. For example, *Eucalyptus amplifolia* Naud. has been coppiced cultured through four rotations in northern Florida with annual yields reaching 23 dry tons per hectare. Somewhat similar high yields have been reported for *Eucalyptus saligna* Sm. in Hawaii. In each case, the key was the careful identification and selection of genetic material (species and seed sources) appropriate for given site conditions and the application of compatible management techniques.

It is apparent from a number of current field tests that the full potential of this genus has not been exploited and many more seed source (provenance) investigations need to be initiated within a narrow range of growing conditions. While it is very unlikely that true cold-hardy Eucalypts will be found, there are many situations in which this genus will be useful. At this time, only approximately half of the species have been tested to any degree in the United States. There are some subgroups of the Eucalypts in which virtually little is known, even in Australia where it is native. What is needed is a better coordinated effort and a better understanding of the considerable variation in this genus.

Paulownia Tomentosa

Although *Paulownia tomentosa* (Thunb.) Sieb. & Zucc. was introduced into the United States in about 1845, virtually little is known about other species of this genus in the United States, nor the extent of genetic variation in *Paulownia tomentosa*, the most northern species in Asia. Since its first introduction, mainly in the southern United States, *Paulownia tomentosa* is now found growing wild in many parts of the United States, including the western U.S. and even in the northeast as well as in the south. In China, its source of origin, yields of 36 to 53 cubic meters per hectare have been reported. It should be noted that this species is not grown for its biomass alone, but for its use as a quality furniture wood.

Many report this species as a weed tree for its poor form and rather large shade leaves. This is not surprising since there has been very little genetic selection of seed source for its adaptability to given sites, nor selection for form or growth rate. In the last 15 years, studies have been initiated to select from *Paulownia tomentosa* populations within the United States for improved growth and form. Still, current efforts to produce *Paulownia* spp. commercially have not been completely successful. In addition to a lack of appropriate plant material (seeds), optimum management techniques have not been applied.

While reviewing forest genetics activities in the People's Republic of China, both natural and planted stands of *Paulownia* spp. were studied. Currently, seven species and six varieties are commonly recognised.

However, the Chinese list nine species of *Paulownia*. Many of the species tend to be subtropical and are not suitable for the continental United States. In 1986, four provenances each of *Paulownia elongate* (S.Y. Hu), *P. fargessii* (Franch), *P. fortunei* (Seem) Hemsl. and *P. tomentosa* collected in China were planted near Birmingham, Alabama; Tillman, South Carolina; Morganton, North Carolina; and the Garrett-Allegheny County area of Maryland. These species and provenances were selected as being good candidates for the southern and southeastern United States.

On the whole, seedling survival after 2 years has been good with height growth ranging from 0.5 to 8 meters. It is too early to make any conclusion but it does appear that other *Paulownia* species can be grown in the United States. At 5 years, we will have an opportunity to evaluate form as well as height growth. As a comparison, *P. elongate* in a park in Peking had a height of 10.6 meters at age 6 while a roadside tree of *P. fortunei* near Kweilin, Kuangsi, had a height of 20.8 meters at age 11.

From the experience in China, *Paulownia* spp. prefer warm climates with 24°C to 30°C being the most favorable. *P. tomentosa* can withstand temperatures as low as -20°C, and *P. elongata* -15°C. Both *P. fortunei* and *P. fargessii* appear to be less cold hardy than *P. elongate*. Cold hardiness will need to be further determined under a number of growing conditions. Additional studies will be needed to determine appropriate management techniques for this genus. Yet, its rapid growth, attractive flower, and excellent wood quality make it a genus that needs to be considered for further use in the United States.

Black Locust

Black Locust (*Robinia pseudoacacia* L.) is a nitrogen-fixing legume, native to southeastern North America and now naturalised extensively in the temperate regions of North America, Europe, and Asia. This tree is useful but underutilised for lumber, poles, wood fiber, land reclamation, beekeeping, fuel and forage. It grows very rapidly survives droughts and severe winters, tolerates infertile and acidic soils, and produces livestock feed nutritionally equivalent to alfalfa (*Medicago sativa* L.). Commercial feed production, as silage, hay or leaf meal, appears feasible.

The primary use of black locust wood has been for fence posts which, due to flavonoids in the heartwood, can endure for over 100 years in the soil. The lumber is one of the heaviest and hardest in North America, but the supply is limited due to damage from the locust stem borer, *Megacyllene robinae*. In Hungary, where the insect is not a problem, black locust is a major timber tree (Fig. 5).

Figure 5. Black locust

The wood is suitable for chemical pulping systems in the paper industry, and small amounts are presently being used in Michigan, where the Menasha Corporation established a test plantation in 1988. Widely used for erosion control and reforestation, black locust is ideally suited for woody biomass plantings, and commercial energy production may eventually become one of its primary uses in the U.S. Its virtues include nitrogen fixing ability, inexpensive propagation by seed, rapid vegetative propagation, adaptability to a wide range of site conditions, rapid juvenile growth, high heat content of the wood, and prolific regrowth after cutting.

Flowers

In Hungary, black locust plantings are vital to commercial honey production. The showy white pea-like flowers are borne on new growth three to four weeks after growth begins. Vigorous specimens can bloom in three years from seed, much earlier than most trees.

Leaves

The leaves are used for livestock feed in the Republic of Korea and in Bulgaria. In the highlands of Nepal and northern India, where black locust is naturalised, it is an important fodder tree. Branches above the reach of livestock are cut when other green for-ages are scarce, and the wood is used later for fuel. Ground black locust tops including woody stems, from first and second harvests, were found to be comparable to alfalfa (*Medicago sativa* L.) with 23-24% crude protein, 7% lignin, and 4.2 kcal/g. Ruminal digestion by cattle was also equivalent. When planted at close spacings, the new growth can be harvested with conventional farm machinery for silage or hay. The compound leaves can be separated and ground for a high-protein ingredient of commercial feeds. Because black locust thrives on sites too marginal for alfalfa, it merits further study as a forage crop.

Leucaena Hybrids

Leucaenas (genus *Leucaena*, Mimosoideae-Leguminosae) are among the most versatile of trees. Few tropical trees have matched leucaena's ability to provide quality fuelwood, fodder, polewood, green manure, shade, erosion control and other useful products in the tropics and subtropics. The N-fixing capability of these legumes is also of importance, particularly in low-input farming systems. Currently 2-5 million hectares worldwide are estimated to be planted to leucaenas.

Natural evolution has created impressive variability in this genus, but the majority of leucaena production has until recently been limited to *L. leucocephala*, the only species distributed pantropically. Both its shrubby and arboreal forms are high-yielding and coppice readily. Both tend to produce heavy seed crops, making them weedy and diverting fixed carbon from the desired wood or fodder. Additionally, this species is often heavily damaged by a psyllid insect.

Artificial hybridisation has greatly expanded the way farmers can exploit these multipurpose trees by creating species hybrids with unique growth habits, ecological adaptations, wood and fodder quality, pest resistance and high productivity. The most significant of these hybrids are those which are seedless and have high psyllid resistance.

Leucaena Bentham has had more than 50 species ascribed to it, but most are viewed as synonyms. Thirteen species are widely recognised. Taxonomic distinctions between the species are strongly supported by geographic and ecological distributions, diverse tree and pollen morphology (e.g., leaflets per leaf range from 30-13,000+) as well as chromosome counts. New taxa can be expected; the most recently validated species is *L. salvadorensis* Standley, rediscovered by Hughes. Natural hybrids have been only infrequently discovered, evidently due to geographic and phenological isolation.

Leucaena species are native to the neotropics from Texas to Peru. They have colonised such diverse regions as Alpine, Texas, which has three months of snow during winter, and Panamanian rainforests. They range from hot and Mexican and Central American lowlands to the highlands (2000-3000 in elevation). Some species grow as low (5 m), highly forked shrubs while others grow as trees up to 20 m in height with 80 cm diameter at breast height (dbh). All are marked by high-quality foliage attractive to herbivores and by dense wood favored for fuel. Post-Colombian introductions of cattle and goats decimated many natural stands in regions which were dominated by leucaenas as evidenced by landmarks, villages, mountains and states (e.g., Oaxaca, Mexico) which were named after leucaena.

Several seed collection trips by Brewbaker and colleagues resulted in almost a thousand collections of leucaena species being grown in Hawaii since 1962. Leucaena seed collection programs of Oxford Forestry Institute (OFI) in England and the Commonwealth Scientific and Industrial Research Organisation (CSIRO) in Australia have expanded this world collection. The Nitrogen Fixing Tree Association founded and promotes the International Leucaena Trial (ILT) program of varietal and management trials, and publishes the annual journal *Leucaena Research Reports*.

L. Leucocephala

L. leucocephala is pantropical, the result of distribution via Spanish galleons from Mexico to Southeast Asia in the early 16th century. Leucaenas served as fodder and bedding for the animals which the Spaniards shipped. Unfortunately only the shrubby strain of *L. leucocephala* was involved. This "common" form seeded abundantly and aggressively colonised much of the tropics, notably on sub-humid alkaline soils, especially coralline islands. By the late 19th century, its value as a shade crop for the new coffee and cacao plantations of Asia promoted further international distribution and planting.

For centuries, Mexican and Central American naives supplemented their diets with protein-rich seeds from young pods of at least four *Leucaena* species. Seeds were eaten fresh or boiled. *L. esculenta,* named for its large mild-tasting seeds, is the best known species for this use, and is commonly grown in Mexican midlands today for its seeds and as a green manure crop. *L. esculenta* seeds heavily in these regions, but has not produced seed well in Hawaii or Taiwan. Other species whose seeds are eaten include *L. leucocephala, L. macrophylla* and *L. pallida.* Leucaena issues contain 1-5% mimosine, a thermolabile amino acid which is readily destroyed by heat treatment. A degradation product, dihydroxypyridine, also has toxic properties.

Leucaenas were associated with most pre-Columbian Indian civilisations located between Honduras and Southern Mexico, which all depended on maize as their main food staple. During fallow periods, legume trees could have restored fertility rapidly to maize-cropped soils in areas such as the Yucatan peninsula. The regularity of maize-based civilisation collapse (Olmec, Maya, Zapotec, Teotihuacan, Anasazi) probably related to co-evolved maize diseases or pests (Brewbaker 1979), rather than to soil depletion. Leucaenas are found associated with the ruins of most Mayan ceremonial centers today. Were they recognised and used as green manure crops, as well as sources of wood for construction and fuel? The high response of maize to applied nitrogen, organic or inorganic, is known to all maize farmers, and would hardly have eluded the sophisticated maize growers among Maya or Anasazi.

Interspecific Hybridisation

We have been attempting since 1981 to produce seed via hand-pollination of all possible species hybrids among thirteen *Leucaena* species (182 combinations, including diploid and tetraploid *L. diversifolia*). One hundred and seventy-one (94%) of these species combinations have been tested, and 148 (87%) involved crosses of at least seven inflorescences (average 16.1 florets pollinated per inflorescence). A total of 2941 inflorescences (47,102 florets) have been hand cross-pollinated. Self-compatible species [*L. diversifolia* (4x) and *L. leucocephala*] were hand-emasculated between 3 and 5 AM on the morning of anthesis.

Hybrid verification was morphological and/or chromosomal. The genus is highly intercompatible (46%), with 79 combinations producing F_1 interspecific seed, 15 producing abortive seed and 76 incompatible combinations. Fifty-seven of the 78 combinations (74%) producing apparently-viable seeds were grown and verified; in addition one more hybrid was selected among open pollinated progenies: *L. macrophylla* x *L. diversifolia* (2*x*). Unverified hybrid combinations either did not germinate because of damage from seed beetles (*Araecerus levipennis* Jordan) or died a few days after germination, before verification was possible. Interspecific compatibility in the genus appears likely to increase with further testing to nearly 70%.

Fertility and seediness

Twenty-one hybrids (38%) have not yet flowered, some of which were stunted. Twenty-six hybrids (45%) produced seeds via open pollination; average seed production of these was estimated to be half that of *L. leucocephala*. Only thirteen hybrids (22%) have flowered without setting any pods. Of these, six have flowered well and are represented by numerous trees to show they are seedless. The relatively low proportion of seed sterile hybrids is probably related to the buffering effect of high chromosome numbers in the genus. Eight of the thirteen seedless hybrids (62%) are triploids, four are diploid (31 %) and one is an unusual tetraploid produced from $2x$-$4x$ mating, *L. retusa* x *L. diversifolia* (F_1 is $2n = 4x = 108$).

Recent worldwide damage to *L. leucocephala* by psyllids attests to the probable narrow gene base of this species, and we treat our hybrid program largely as a way to broaden the gene base of *L. leucocephala*. Because it is tetraploid ($2n = 104$), as are other high-yielding species *L. diversifolia* and *L. pallida*, our breeding has largely been at the tetraploid level. These three tetraploid species produce excellent single-cross hybrids in all combinations, and open-pollinated F_2s from these planted in multinational sites have generated much interest.

Three-way hybrids among the tetraploid species have shown promise for broadening the gene base from which breeders can make selections. Some of Sorensson's three-way hybrids were found to be self-compatible (the female parent was self-incompatible), allowing a single seed descent breeding approach. Tetraploid three-way hybrids were also derived through unreduced gametes between two tetraploid species and the frost-tolerant diploid species *L. retusa*. The use of unreduced gametes and of colchicine for chromosome doubling may facilitate gene transfer from diploid species to *L. leucocephala*.

Adaptability and use

Many tropical environments are hostile to *L. leucocephala*. These include regions above 500-1000 m elevation with mean annual temperatures below 22°C, areas where temperatures remain below freezing for more than a few hours at a time, and sites with acid and/or high aluminum soils below pH 5. Interspecific hybrids afford possible solutions to these problems. For example, hybrids of *L. diversifolia* (4x) x *L. leucocephala* averaged 4.5 m/year/height increment in a two-year period at Waimea, Hawaii (850 m elevation, mean annual temperature 17°C). Two- and three-way hybrids involving *L. leucocephala* and frost-tolerant *L. retusa* are being checked for frost resistance.

Hutton tested several species and hybrids for acid soil tolerance in Colombia and Brazil, and *L. diversifolia, L. lanceolata* and *L. shannoni* (all $2n = 52$) showed potential as parents. His current program is largely based on hybrids derived from triploid *L. diversifolia* x *L. leucocephala* (F1 is $2n = 3x = 78$).

Gum production

Legumes produce the major gums used in foods and other industries. Gum arabic from *Acacia senegal is* particularly well known. *L. leucocephala* has infrequently produced a translucent tan gum under stress. Gummosis was first observed in India following attack by *Fusarium semitectum*, and was later seen in Hawaii following attacks by *Phytophthora drechsleri* and wood-boring beetles. Gum production was sporadic, low yielding, and was often associated with wood dieback Analysis of several leucaena gums has revealed that they have the closest match to gum arabic of any gums tested from a hundred or so tropical trees. Although toxicity and related studies are needed, leucaena gum may have potential for use as a substitute for gum arabic (Anderson 1986).

L. leucocephala x *L. esculenta* hybrids segregated trees which exuded gum copiously, and have not had wood dieback. Nine hybrid trees of this pedigree were grown at Waimanalo, Hawaii for four years during which approximately a third failed to produce gum, another third exuded gums sporadically and another third exuded gums heavily (Fig. 5). These high gum yielders exceeded the mean annual per-tree gum production (250 g/tree) of gum arabic by *Acacia senegal.* Gum production appeared as balls or drippings from mature bark, and was heaviest in the dry season. Hybrids of this type are seedless, have good vigor and psyllid resistance, and could prove promising for gum production

Furniture, construction timber and polewood

Like the arboreal forms of *L. leucocephala,* a number of species hybrids appear to produce the thick straight boles required for use in furniture or construction timber. *L. pulverulenta* x *L. leucocephala* was preferred in Indonesia about 50 years ago for its straight boles, and fast growth at cooler upland sites. Early studies in Hawaii showed it to be a vigorous hybrid, although its psyllid susceptibility now curtails its use in lowland sites. Hybrids which have potential to form trees with dbh of 30 cm and straight boles of 5 m include the following:

Seed parent		Pollen parent	Chromosome number of F_1
L. leucocephala	x	*L. pallida*	104
L. leucocephala	x	*L. diversifolia*	104
L. pulverulenta	x	*L. diversifolia*	80
L. diversifolia (2x)	x	*L. diversifolia* (4x)	78
L. diversifolia	x	*L. leucocephala*	78

Silvicultural practice should include dense planting (e.g., 10,000 stems/ha) with thinning for fuelwood at one and three years, and harvest after six to eight years. Polewood is commonly used in the production of vine crops (black pepper, passion fruit, pole beans) where long, straight, thin poles are preferred. The hybrid *L. diversifolia* (4x) x *L. pallida* is psyllid resistant and grows as a pseudo-shrub with many long straight branches. High-density planting of the five hybrids listed above would also produce stems suitable for polewood.

Fodder production

Fresh herbage yields of *L. leucocephala* (40-80 fresh t/ha-yr) matched or-exceeded those of other tropical legumes when moisture was not limiting. Psyllid resistance of hybrids like *L. leucocephala* x *L. pallida*, however, exceeds that of any *L. leucocephala*, permitting higher fodder yields under psyllid attack. *L. leucocephala* x *L. pulverulenta* and *L. diversifolia* (4x) x *L. leucocephala* (R.A. Bray personal comm.) both have outyielded *L. leucocephala* cut for fodder under certain conditions.

Short heavily forked trees are preferred for herbivore browsing. Most accessions of the tetraploid species *L. pallida* are low forking and they confer this trait to some of their hybrids with *L. leucocephala*, some low shrubby dwarfs also result. Low mimosine contents (0.5-1.0%) would enable leucaena fodder to be fed in higher quantities to nonruminants like chickens, horses and tilapia. Examples of *Leucaena* species and hybrids with low mimosine are *L. diversifolia* and *L. pulverulenta* and their hybrids.

Shade and support

An important use of leucaenas is as a shade or nurse tree in plantations of coffee, cacao, quinine or tea, or as supports for vine crops. *L. leucocephala's* seediness is a major deterrent to this use, as its weediness raises management costs. Some coffee and tea plantations in Indonesia graft seedless leucaena clones for use as nurse crops. Our pollen analysis of two of their clones suggests they are aneuploids derived from triploid species hybrids of *L. diversifolia* (2x) x *L. leucocephala* or *L. pulverulenta* x *L. leucocephala*.

Uses of species hybrids

Non-seedy or seedless clones are attractive options for most of leucaena's wood uses, including fuelwood, pulpwood, roundwood, charcoal, parquet, and craftwood. Hybrids such as *L. retusa* x *L. esculenta make* attractive home ornamentals. Vegetatively propagated clones of self-incompatible species which are attractive to honeybees, like *L. lanceolata* and *L. shannoni,* should have longer flowering seasons due to inhibition of seed set.

Seedless Hybrids Production

Commercialisation of seedless leucaenas requires economic methods of production by vegetative or seed propagules. Rooting of vegetative cuttings of leucaenas has only been successful in temperate greenhouses. Mericlone and other tissue culture techniques have similarly succeeded only under experimental laboratory conditions. Grafting has been used with some success in cooler regions of Indonesia. These methods must be adapted for large-scale use. A second method of producing seedless leucaenas exploits the self-incompatibility (SI) characteristic of all diploid species and of *L. pallida*.

A SI species would be cloned and used as a female to hybridize with SC species. Species would be interplanted at an appropriate ratio of seed to pollen-parent and allowed to produce interspecific hybrid seed through open-pollination. Promising hybrids that lend themselves to this approach include the triploid hybrids *L. diversifolia* (2x) x *L. leucocephala,* and *L. pulverulenta* x *L. diversifolia* (4x).

A third method to produce seedless leucaenas could involve the self-incompatibility system. Inbreds would be produced through "pseudo-self-fertility" treatments not yet applied to leucaena, and S allele homozygotes identified. When two such homozygotes are planted in isolation, all seeds produced by both parents are of a single *S* allele heterozygote, e.g., *S1S2*. When these seeds are grown in isolation as a plantation, seedless progeny will result. Gamete sterility of the triploid type is not required, thus expanding the species and species hybrids that could be exploited commercially. Marketing of hybrid seeds is a prerequisite for a successful leucaena seed industry. *L. leucocephala* and other species have not attracted seed industries because heavy and early seeding limits profitability.

4

Crop Physiology

Grain Position and Micronutrient Concentrations in Wheat

Plant breeding and improved crop management techniques successfully increased grain yield of bread wheat and, in so doing, effectively reduced food shortages and malnutrition during the twentieth century. The future poses a similar challenge: by 2020, the demand for wheat is expected to be 40% greater than its current level of 552 million tons per year. Now, however, a unique opportunity exists for agriculture to invest in developing more nutrient-dense staple food crops that could help reduce not only energy but also nutrient malnutrition.

Welch and Graham proposed a new balanced nutrition paradigm for crop production. They point out that although the world's food supply in recent years has been sufficient, it does not promote an adequate nutritional balance. Breeding staple crops with high nutrient concentration in the grain has been proposed as a low cost, sustainable strategy for reducing mineral deficiencies in humans (more than 2 billion people suffer from deficiencies worldwide) and crops. To explore this possibility, the International Maize and Wheat

Improvement Center (CIMMYT) and other international centers and research institutions have been participating in the Consultative Group on International Agricultural Research (CGIAR) micronutrient project since 1995.

The project is studying the feasibility of developing micronutrient-dense cultivars of wheat, rice (Oryza sativa L.), maize (Zea mays L.), phaseolus beans (Phaseolus vulgaris L.), and cassava (Manihot esculenta Crantz). Its objectives are to determine: (i) available genetic variability that could be exploited in future breeding programs; (ii) the genetics and physiology of the selected traits; and (iii) the bioavailability of Fe and Zn in the best wheat selections. A better understanding of nutrient physiology could help design wheat breeding strategies aimed at increasing grain yield without reducing grain nutrient content. Nutrient-rich grain will, in turn, ensure a balanced nutrition in consumers, especially those who rely on wheat for a major portion of their calorie intake.

Wheat breeding has produced significant reductions in N and P concentrations in the grain due to biomass dilution. Reductions in grain N concentrations as a result of plant breeding have also been reported for maize and sunflower (Helianthus annuus L.). Therefore, it could be expected that nutrient concentration in individual grains has been affected similarly by dilution, that is, the heavier the grain the greater the dilution effect. Individual grains of modern wheat cultivars, however, have shown significant variation in N concentration depending on their position within the spike.

Simmons and Moss and Herzog and Stamp found that lighter grains (mainly those in distal positions on the spikelet) had lower N concentrations than heavier grains (generally in proximal positions on basal-central spikelets), which suggests that grain position can have considerable impact on nutrient concentration. Therefore, future breeding strategies aimed at increasing grain yield while maintaining or increasing grain nutrient concentrations should take into account both the biomass dilution effect and nutrient distribution within the spike.

In the past, wheat yield gains have been associated with increases in the number of grain rather than in grain size. Increases in grain number have been due mainly to improved grain set in more distal positions of the spikelet. Miralles and Slafer showed there was a greater proportion of grains in distal positions in semidwarf wheat than in standard-height isogenic lines. The weight of these distal grains was clearly lower than that of grains from florets more proximal to the rachis. This trend suggests that continued selection for additional grains at more distal positions on the spikelets without parallel selection for higher nutrient concentration will result in continued reductions in nutritional quality and a greater imbalance in nutrient concentrations among grains. The effect of grain position on N concentration has been evaluated, but the manner and extent to which the position within the spikelet affects grain concentration of other macronutrients have not been reported.

The finding that grain N concentration is variable within the spike supports the hypothesis that grain concentration of other macronutrients (e.g., P) is also affected by grain position. Ortiz-Monasterio and Graham found a very weak association ($r^2 < 0.33$) between Fe and Zn grain concentrations and year of release of CIMMYT cultivars, and a strong ($r^2 = 0.72$) linear relationship between grain yield and year of release in the same set of cultivars.

It has been proposed that dry matter distribution influences N partitioning within the spike, since distal grains achieve less weight than those located closer to the rachis. At the same time, the dry matter and water dynamics of growing grains are closely related in wheat cultivars. Little is known, however, about the relationship between dry matter distribution and macro- and micronutrient concentration in grains in different positions and how they accumulate during grain filling. Synthetic hexaploid wheats (T. durum x Aegilops tauschii) can be used to study these relationships, since they have shown differences in dry matter distribution compared with modern cultivars. Synthetic wheats may also be a source of variability for grain nutrient concentration, given that A. tauschii, a parent of the synthetics, has shown higher grain concentrations of micronutrients than T. aestivum.

Seed and Oil Concentration on Seed Filling Period

Weight per seed and seed oil concentration in sunflower are determined during the seed filling period (from end of flowering to physiological maturity). Weight per seed is determined during the entire period. Seed oil concentration begins to increase rapidly a few DAF and continues to increase until a few days before physiological maturity, when it stabilizes. Changes in environmental conditions during the seed filling period could potentially affect both yield components. Relationships between yield components and environmental conditions have been established for several crops. In sunflower crops grown under good moisture and nutritional conditions, seed number per unit land area was closely correlated to photothermal quotient [daily average solar radiation/(daily average temperature - base temperature)] from floret initiation to seed set stage.

Recently, weight per seed was closely related to intercepted PAR from the end of flowering (R6 + 46°C d, base temperature = 6°C) to physiological maturity in two hybrids, one with high and the other with low seed oil concentration. The variation in oil concentration in the hybrid with high seed oil concentration was also related to intercepted PAR during the same period ($r2$ = 0.93). The duration of the R6 to R9 period can be highly variable as the time to physiological maturity varies among years in responses to environmental conditions.

Therefore, relationships between intercepted PAR during R6 to R9 and weight per seed and oil concentration could then only be used a posteriori to describe rather than predict seed weight or oil concentration, or need to be coupled with more complicated models to estimate the date of physiological maturity. To achieve a fast and simple prediction of both oil yield components, it could then be useful to establish relationships in which intercepted PAR is accumulated between fixed crop stages or fixed dates in time. Assimilates used in grain filling mainly originate in current photosynthesis.

During the seed filling period, weight per seed shows a phase of fast increase during which sink demand for assimilates is strong. A phase of fast oil concentration increase has also been observed. Therefore, a developmental interval during which final weight per seed and oil concentration are closely linked to intercepted PAR could exist. Determination of this critical developmental interval and of quantitative relationships between weight per seed and oil concentration and intercepted PAR per plant could be useful to improve sunflower crop management and modeling yield so that crops could be managed to maximize available environmental resources during such an interval.

Genotypic Variation for Glycinebetaine in Sorghum

The breeding of crop plants for environmental stress tolerance has been difficult and slow. The quantitative nature of stress tolerance and the problems associated with developing appropriate and replicable testing environments make it difficult to distinguish stress-tolerant lines from sensitive lines. One approach to a better understanding of plant stress tolerance is to isolate those characteristics that are proposed to contribute to stress tolerance and to determine their relative importance. Only then can focused breeding approaches be developed.

Accumulation of solutes, either actively or passively, is an important adaptation mechanism for plants in response to osmotic stress. Cellular dehydration is a general consequence of osmotic stresses, including water deficit and high salinity levels. In response to this condition, many organisms synthesize solutes that either help retain water within cells or protect cellular components from injury caused by dehydration. One such solute, glycinebetaine (GB) has been shown to act in both capacities.

Glycinebetaine is thought to play an important role in plant adaptation to saline and arid environments. Genes determining GB accumulation are thus of considerable interest in plant breeding for stress environments. Many cereal crops accumulate GB, although some, notably rice (*Oryza sativa* L.), do not. Glycinebetaine-nonaccumulating genotypes of both sorghum and maize have been identified, although in sorghum, lines that do not accumulate GB are rare. The relative concentrations of GB vary both among and within species. The levels of GB found in sorghum are as much as 10-fold higher than those observed in maize, and within maize GB-accumulating lines, a wide range of relative levels exists.

GB nonaccumulation appears to be caused by recessive alleles of a single locus in maize that results in an inability to convert choline to betaine aldehyde, the first committed step in the synthesis of GB. At least 13 diverse GB-nonaccumulating maize genotypes are allelic with respect to this locus. Near-isogenic maize lines that accumulate GB are more salt tolerant than their GB-nonaccumulating sister lines, further supporting the suggestion that GB accumulation is an important stress tolerance mechanism in plants.

GB is synthesized in plants from serine via ethanolamine, choline, and betaine aldehyde. Although other pathways may exist (such as direct N-methylation of glycine), the pathway from choline to GB is the only one that has been identified in GB-accumulating plant species to date. Choline monooxygenase (CMO), which catalyzes the oxidation of choline to betaine aldehyde, and betaine aldehyde dehydrogenase (BADH), which oxidizes betaine aldehyde to GB, are both induced under osmotic stress.

Since GB accumulates in sorghum in response to salinity stress, it was of interest to determine the range and extent of variability for this trait among diverse genotypes of this species. GB represents the major QAC in sorghum, as it does in maize, and that this QAC is genetically and environmentally regulated.

Carbon Isotope Discrimination

Drought, defined as water deficit, and often combined with high temperature stress, is one of the greatest constraints to cereal grain yield in Mediterranean areas. For crops such as durum wheat, grown under rainfed conditions, agricultural practices are not sufficient to mitigate the effect of drought, and plant breeding has become the best tool for yield increases. Carbon isotope discrimination (Δ), when measured in plant dry matter, integrates transpiration efficiency, the ratio of net photosynthesis to water transpired, over the period during which the dry matter is assimilated, and transpiration efficiency being negatively related.

Although water status during growth dramatically affects yield and, a specific environmental variable responsible for the positive relationship usually found between and yield across growing conditions (i.e., trials) has not been unequivocally identified. For breeding purposes, it is crucial to assess how the growing environment affects the relationship between and yield across genotypes.

Patterns in this relationship may be masked by phenological differences among genotypes that may affect yield and also , especially in drought-prone environments. For example, under Mediterranean conditions those genotypes with fewer days from sowing to heading or to anthesis show higher values probably because they attain grain filling with more water in the soil, whereas the evapotranspirative demand is lower.

A total of 12 field trials were performed from 1997 to 1999 in two Mediterranean regions, northeastern (NE) and southeastern (SE) Spain, and under two contrasting water regimes (rainfed and irrigated) within each region. Presowing subsoil moisture was not relevant for any of the trials assayed.

Trials consisted of 25 durum wheat genotypes grown in randomized complete blocks with four replicates in plots of 12 m^2 (six rows, 0.20 m apart). These genotypes included four commercial Spanish cultivars (Altar-aos, Jabato, Mexa, and Vitrón), and 21 genotypes from the eastern Mediterranean basin, including commercial cultivars and advanced lines of the CIMMYT/ICARDA durum wheat breeding program (Awalbit, Bicrecham-1, Chacan, Chahra-1, Haurani, Korifla, Krs/Haucan, Lagost-3, Lahn/Haucan, Massara-1, Moulchahba-1, Mousabil-2, Omlahn-3, Omrabi-3, Omruf-3, Quadalete//Erp/Mal, Sebah, Stojocri-3, Waha, Zeina-1, and Zeina-2).

The genotypes were chosen to represent a wide range of genetic variability in terms of agronomical characteristics. Moreover, to minimize the potential interference of phenology on yield and Δ, these genotypes were selected with a relatively narrow range of variability in the number of days from planting to heading, days to anthesis, and days to physiological maturity (a mean range of 7 and 6 d for heading and anthesis dates, respectively).

The times from sowing to heading (DH) and anthesis (DA) were recorded when more than half the plants in a plot reached the stages 55 and 65 of Zadoks' scale, respectively. Thermal time was calculated in growing degree-days (GDD) by summing the daily values of mean temperature, with a base temperature of 0°C. Physiological maturity was recorded when most of the plants had reached Zadoks' stage 87, representing kernels in hard dough stage. Plots were harvested mechanically when the grain was ripe, yield (kg ha^{-1}) expressed at a 100 g kg^{-1} moisture content.

Environmental Parameters

For each trial, W_i was calculated as the sum of rainfall and irrigation (when applicable) for the following periods: sowing to heading, sowing to anthesis, sowing to maturity, heading to anthesis, heading to one week after anthesis, heading to maturity, anthesis to maturity (grain filling), and from one week after anthesis to maturity. Mean temperature and reference ET_0 for the same periods were also assessed. ET_0 was calculated from the mean maximal and minimal average temperatures of each period using the PC-program ET_0 (Sub-zero Evapotranspiration) version 1.04.

For each plot, a sample of about 2 g of mature kernels was oven dried and finely ground (mesh diameter of 0.5 mm). The $^{13}C/^{12}C$ ratio of samples was subsequently determined by mass spectrometry at the Serveis Científico-Tècnics de la Universitat de Barcelona, Spain. Samples of 0.7 to 0.9 mg were combusted in an elemental analyzer and the $^{13}C/^{12}C$ ratio was measured with an isotope ratio mass spectrometer (Delta C, Finnigan Mat, Bremen, Germany) operated in continuous flow mode. A system check for elemental analyses was achieved with an interspersed working standard of atropine. Stable carbon isotope composition was expressed as ^{13}C values, where $\delta^{13}C(\%)$= [(R sample/R standard)-1] x 1000, and R is the $^{13}C/^{12}C$ ratio.

Secondary standards of graphite, sucrose, and polyethylene foil calibrated against Peedee belemnite (PDB) carbonate were used for comparison. The accuracy of the 13C measurements was ±0.1%. Following Δ Farquhar et al., was further calculated from ^{13}C as $\Delta = (\delta_a - \delta_p)/(1 + p)$, where a and p refer to air and plant, respectively. On the PDB scale, free atmospheric CO_2, δ_a, has a current composition of approximately -8%.

An analysis of variance was performed for grain yield and Δ, considering days from sowing to heading or from sowing to anthesis as the covariant. Means of trials for yield and Δ were compared by the LSD test at P = 0.05. Stepwise discriminant analysis was used to ascertain the relationship between either Δ or grain yield as the dependent variable and three different periods of the crop cycle (sowing to heading, heading to anthesis and anthesis to maturity) for each environmental parameter as the independent variables.

5

Crop Breeding

Inheritance and Expression of Transgenes

The transgene inheritance and phenotypic expression will assist in the development of efficient breeding strategies for transgenic germplasm, and guide research into the improvement of transformation techniques. Transgenic Loci introduced into higher plant species frequently display unpredictable patterns of inheritance and expression. Transgene instabilities—either of inheritance or expression—complicate the identification, selection, and use of transgenic lines. The development of efficient breeding strategies, and of improved transformation systems, should be based on knowledge acquired from empirical assessments of the inheritance and expression patterns of transgenic loci.

Studies of transgene silencing and inheritance have shown that unpredictable behaviour is common. Variable transgene expression among clonal cell lines of mice (Mus musculus) has been reported and compared with the position effect variegation phenomenon described for yeast (Saccharomyces cerevisiae) and Drosophila; it is reasonable to assume that a similar phenomenon could occur in transgenic plants and contribute to irregular transgene expression. Inheritance patterns of transgenes also exhibit irregular behaviour, and frequently deviate from Mendelian expectations. For example, Srivastava et al. observed the loss of transgenes in R_3 wheat plants following loss of expression in prior generations, and poor transmission of integrated transgenes to progeny in some transgenic lines has been observed also in maize (Zea mays L.) and barley.

Features and Use of Transgenic Barley Plants

Transgenic barley lines were derived from seven independent transformation events which were produced and described by Wan and Lemaux. All lines were derived from phosphinothricin acetyltransferase (PAT)-positive callus lines which were shown by Southern hybridizations to have incorporated bar and pRsa1BYDVcp into high molecular weight genomic DNA.

In transgenic families; each family was comprised of progeny from a single transformation event. Each family was further subdivided into one to four T_0–derived lines, each comprised of all progeny of a particular T_0 plant. In several cases, additional within-line selections were made, resulting in several T_3– or T_4–derived sublines for these selected lines.

Six families contained the coat protein gene from a PAV isolate of the BYDV driven by the cauliflower mosaic virus (CaMV) 35S promoter, and the bar and uidA genes driven by the maize ubiquitin promoter and first intron encoding, respectively, resistance to the herbicide glufosinate-ammonium and β-glucuronidase activity. An additional family, GP717B-9, differed in that it did not contain the coat protein gene. The agronomic performance of families derived from each of the transformation events has been described.

Transgenes Expression from the T_1 to the T_3

Each T_1 line used in these studies was produced by self-pollination of individual T_0 plants, as described in Wan and Lemaux. For each transgenic line, 18-30 T_1 plants were grown in the greenhouse during the winter season of 1993–1994 and harvested in bulk to produce T_2 seed that was used for field studies in 1994; T_3 populations were produced by bulking equal numbers of seeds from each of 30–40 T_2 plants grown under irrigation at Aberdeen and Burley, ID.

Progeny from each line were grown in Speedling 128-cell flats (Speedling, Inc., Sun City, FL, USA) in a potting mix of sand:peat:vermiculite (3:2:2) supplemented with commercially available slow-release fertilizer. All plants were grown in greenhouses with natural light. Temperatures ranged from 10 to 25°C. The expression of bar was assayed by applying an herbicide solution [0.5 g kg^{-1} (0.05%) a.i. glufosinate-ammonium; ammonium-DL-homoalanin-4-ylmethylphosphinate; to sections of two leaves of seedlings at the three- to five-leaf stage, and visually assessing herbicide damage.

Each line was characterized by scoring progeny in two separate tests. Resistant plants were selected at random for generation advance, and transplanted to larger pots containing the same potting mix as described above. Several of these selected plants were subsequently used as parents for various crosses. Polymerase chain reaction (PCR) assays were used to detect the coat protein (CP) and bar genes in selected T_1 and T_3 plants. Seedling leaf tissue samples were collected from all progeny grown for the second test of herbicide resistance, from a negative control (nontransgenic 'Golden Promise'), and from positive controls.

DNA was extracted by the following steps: small (50-mg) samples of seedling tissues were ground in 1.5 µL microcentrifuge tubes suspended in liquid N; on ice, 300 µL of a CTAB-based extraction buffer was added to each sample and mixed vigorously with the ground tissue, followed by the addition of 300 µL of chloroform, and incubation at 65°C for 30 to 45 min; at room temperature, 225 µL of the aqueous phase was removed to a clean microcentrifuge tube and DNA was precipitated with an equal volume of isopropanol, pelleted, and rinsed with 100 µL of 70% ethanol; and finally, pellets were dried and redissolved in 30 µL of 10:0.01 Tris-EDTA.

Two microliters of each DNA sample, plus a 200-ng - HindIII standard, was loaded onto a 1% agarose gel and electrophoresed for 3 h at 80 V, stained with ethidium bromide, and examined under UV light to determine relative DNA concentration and quality. Samples in which DNA appeared to be excessively sheared or extremely dilute were discarded, and re-extracted from remnant plant tissues stored at -20°C. All DNA samples found to be of suitable quality were further quantified via Hoechst 33258-dye based fluorometric measurements (using calf thymus DNA at 100 ng µL-1 as a concentration standard), and subsequently diluted to 10 ng µL-1 with 10:0.01 Tris-EDTA.

Polymerase chain reaction assays were conducted using primers specific to sequences within the coding region of each gene (for CP: CP-F, 5'-AGAACAAGTTCGGCCAGTGGT-3' and CP-R, 5'-CCATTAATCGCTTCGGCTCT-3'; for bar: BAR5F, 5'-CATCGAGACAAGCACGGTCAACTTC-3' and BAR1R, 5'-ATATCCGAGCGCCTCGTGCATGCG-3'). Thermal cycler conditions were for detection of CP were: one cycle, 94°C 1 min; 35 cycles, 94°C 30 s–55°C 15 s–72°C 30 s. Thermal cycler conditions for detection of bar were:

seven cycles, 95°C 1 min–65°C 30 s–72°C 2 min; 16 cycles, 95°C 1 min–60°C 30 s–72°C 2 min; 16 cycles, 95°C 1 min–55°C 30 s–72°C 2 min; 1 cycle, 72°C 7 min. Twenty-five µL reactions were set up according to directions supplied by the manufacturer's of the DNA polymerase (Amplitaq DNA polymerase was used for CP reactions, supplied by Perkin-Elmer, Branchburg, NJ, USA; DyNAzyme EXT DNA polymerase was used for bar reactions, supplied by MJ Research, Inc., Waltham, MA, USA).

Approximately 25-ng template DNA was used for each CP assay, and 100-ng template DNA was used for each bar assay. Twenty µL of each reaction was loaded onto a 2% agarose gel and electrophoresed for 3 to 4 h at 80 V. For each CP and bar assay, each DNA sample was assayed at least twice, and data was recorded only for assays in which amplified products were of the expected size [CP = 416 base pair (bp); bar = 343 bp], the positive controls produced products of the expected size, and the negative control produced no product. In cases where PCR assay results were inconclusive, DNA was re-extracted from remnant tissues and additional assays were conducted as necessary to produce interpretable results.

Inheritance and Expression of Transgenes in Progeny

Two transgenic families which had relatively good transmission of transgenes through the T_3 generation, and which showed relatively good stability of bar expression (measured as resistance to 0.5 g kg^{-1} glufosinate-ammonium). Several plants homozygous for bar (based on progeny tests for glufosinate-ammonium resistance and PCR assays) were selected at random from T_3 progeny of GP724B-4-9 (designated T_3 #30, 31, 32, and 37) and T_4 progeny of GP717B-59-11 (T_4 #19). These plants were self-pollinated to produce T_3– and T4–derived sublines, and were used as parents in crosses to nontransgenic breeding lines and cultivars. F_1 progeny of these crosses were then backcrossed to these same breeding lines and cultivars. The recurrent (nontransgenic) parent was used as the female for most crosses and for all backcrosses.

Progeny derived from these crosses were assayed in the F_2, F_3, BC_1F_1, and BC_1F_2 generations for glufosinate-ammonium damage based on visual observations; scoring was qualitative (either resistant or susceptible). For the development of backcross populations, putatively nontransgenic BC_1F_1 plants were eliminated based on glufosinate-ammonium susceptibility; the remaining BC_1F_1 plants were hemizygous for bar and therefore the expected segregation patterns of the unselected BC_1F_2 populations were the same as for the F_2 populations (3:1 resistant:susceptible). Resistance or susceptibility to glufosinate-ammonium was determined by applying a commercial formulation (Finale herbicide, AgrEvo, Montvale, NJ, USA) to greenhouse-grown seedlings at the three- to five-leaf stage, as an aqueous solution at a final concentration of 0.5 g kg-1 (0.05% a.i.) using a hand sprayer equipped with a flat-fan nozzle.

The herbicide solution was applied from two different directions to uniformly and thoroughly wet the leaves. Except as noted in the results, these tests were not replicated. Additional tests were conducted to quantify the level of herbicide resistance in populations consisting of single cross- and backcross-derived plants (in the F_5 and BC_1F_4 generations, respectively), transgenic parent sublines (in the T_7 generation), and nontransgenic (recurrent) parents. These transgenic single cross- and backcross-derived populations had been characterized as homozygous for bar, based on progeny tests for qualitatively-scored resistance to glufosinate-ammonium as described above.

Resistance to glufosinate-ammonium damage was quantified based on the dry weights of seedlings following a defined period of growth after application of glufosinate-ammonium solutions from 0 g kg^{-1} to 10.0 g kg^{-1} (0 to 1%) a.i. Herbicide was applied at the three- to five-leaf stage at a rate equivalent to 635 L ha^{-1} using a specialized spray cabinet capable of accurate and repeatable applications. This rate approximated the rate applied in previous experiments using the hand sprayer. One week after treatment with various rates of glufosinate-ammonium, plants were harvested at the soil level and dried for 24 h at 37°C, and weighed. Data were collected from three replicates; each replicate included eight seedlings of each progeny population, transgenic parent subline, or nontransgenic parent.

Application of 0.5 g kg-1 glufosinate-ammonium to sections of nontransgenic Golden Promise leaves resulted in necrosis of the treated leaf sections within 4 to 6 d. Early symptoms of damage were a very slight chlorosis of the treated tissue, followed by a loss of turgor, complete chlorosis, and eventually total desiccation of at least part of the treated tissue (Fig. 1a). Transgenic plants expressing herbicide resistance, when grown under appropriate environmental conditions, ranged from no detectable symptoms to a slight chlorosis (Fig. 1c).

The tests conducted across a period of several years; during that time it became obvious that stress, such as high temperatures (>30°C) typical of summer greenhouse conditions, resulted in reduced, variable expression of glufosinate-ammonium resistance; such data were discarded. Tests conducted under appropriate conditions, however, were found to be highly reliable, as evidenced by the cosegregation of resistance and positive bar assays, and/or by the results of numerous progeny tests.

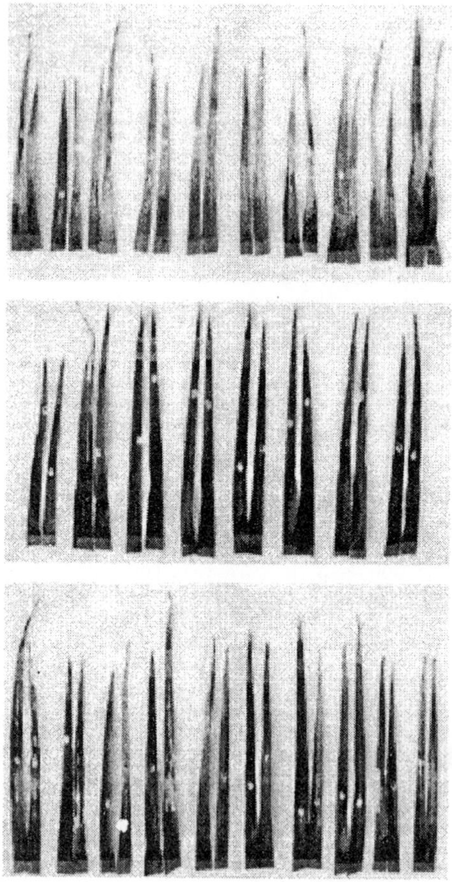

Figure 1. Typical plant reactions to brush application of 0.05% a.i. glufosinate-ammonium to leaf sections of 'Golden Promise' (top), GP724B-4-9 T_3 #30 (middle), and GP724B-4-9 T_3 #31 (bottom). Each pair of leaves is from an individual plant; glufosinate was applied to a 3-cm section of each leaf directly below the punched hole.

Furthermore, particularly for lines with relatively low levels of bar-expression, we have found that lower rates of herbicide application, coupled with daily observation of the development of damage symptoms, are necessary to properly score segregating populations. By carefully controlling the timing and rate of glufosinate-ammonium application, greenhouse conditions, and by making multiple and careful observations of treated plants, we have been able to routinely, and reliably, identify plants expressing bar.

Segregation for herbicide resistance deviated from Mendelian expectations for the segregation of single loci in the majority of transgenic families. No evidence was observed that any of these families had multiple transgenic loci. The segregation patterns of lines within families showed no significant differences, and therefore the data were pooled across lines and presented as family means in Table 1. The level of herbicide resistance appeared to vary among lines, but the nature of these assays did not allow for accurate quantification. Quantification of resistance was examined in subsequent experiments.

Table 1. Transgene inheritance and expression in transgenic lines from seven transgenic T_0-derived families, each from an independent insertion event.

		Glufosinate-ammonium reaction		
Family	No. of T_0-derived lines tested	All tested plants[††],	Plants analyzed,[††]	PCR results,[††]
GP717B-9 T_1	4	142R:42S	none	no data
GP717B-9 T_3	4	73R:152S*	39R:86S*	40+:85-*
GP717B-32 T_1	4	2R:101S*	1R:55S*	5+:51-*
GP717B-32 T_3	4	0R:195S*	0R:99S*	0+:99-*
GP717B-33 T_1	4	75R:65S*	42R:28S*	51+:19-
GP717B-33 T_3	4	25R:191S*	8R:87S*	27+:68-*
GP717B-189 T_1	4	0R:205S*	0R:43S*	32+:11-
GP717B-189 T_3	4	2R:75S*	0R:80S*	44+:36-
GP717B-59 T_1	1	66R:17S	31R:4S*	31+:4-*

Contd...

Contd...

GP717B-59 T_3	1	84R:19S*	30R:2S*	30+:2-*
GP724B-4 T_1	4	182R:84S	106R:18S*§,	109+:15-*§,
GP724B-4 T_3	4	123R:132*	69R:44S*§,	78+:35-*§,
GP724B-96 T_1	4	64R:106S*	51R:55S*	52+:54-*
GP724B-96 T_3	4	11R:222S*	5R:118S*	8+:115-*

* Significant deviations from expected ratios (P = 0.05.

†Expect 3:1 and 9:7 resistant:susceptible or gene (+):(-), T_1 and T_3, respectively.

‡pRsa1BYDVcp (cp) and bar tightly linked; data refers to both genes.

§18 susceptible T_3 plants and 24 susceptible T_3 plants died before DNA samples were obtained. If susceptible plants were transgene (-), segregation ratios would not deviate from expectations.

T_1 progeny showed the expected 3:1 resistant:susceptible (R:S) patterns in three of seven families; none of the families showed the expected 9:7 R:S segregation patterns in the T_3 progeny. These data for segregation for herbicide resistance among T_1 progeny were consistent with preliminary data reported for these families. In all but one case where aberrant segregation was noted, fewer resistant progeny were detected than expected. The exception was for T_3 progeny of GP717B-59 which showed 3:1.

Data from Southern analyses, T_1 and T_3 transgene inheritance patterns, and from F_2 and BC_1F_1 herbicide resistance segregation patterns indicated that there was only one transgenic locus in this family. Additional herbicide assays of T_3 progeny of GP717B-59 were conducted (included in the data presented in Table 1) to verify that sampling error was not the cause of these results. The regeneration of tetraploid transgenic barley plants has been observed to be common, and cytogenetic abnormality was considered as a potential source of unexpected segregation patterns, but limited cytogenetic analyses detected only diploid plants.

Segregation of Transgenes in T_1 and T_3 Progeny

Polymerase chain reaction analyses showed identical segregation patterns for the inheritance of two transgenes, bar and the BYDV CP gene, indicating (as expected) tight linkage. The data on transgene inheritance therefore are presented as single ratios representing both genes in Table 1. The majority of families showed fewer transgenic progeny than expected, indicating poor transgene transmission to progeny.

Two of the families tested (GP717B-189 and GP724B-4) showed expected patterns of transgene inheritance for both generations tested, and GP717B-33 showed expected patterns in the T_1 generation only. Note that for GP724B-4, the assumption was made that susceptible plants that died prior to tissue sampling were transgene negative. Note also the unexpected preponderance of transgene-positive progeny in the family GP717B-59 in both the T_1 and T_3 generations, which is consistent with the data for herbicide resistance.

Comparisons of segregation patterns for transgene expression vs. transgene inheritance suggest failures of transgene inheritance, and not transgene silencing, as the primary reason for the loss of herbicide resistance from this transgenic population. In most families, herbicide-susceptible/transgene-positive progeny were a small fraction of the herbicide susceptible progeny. The most notable exceptions were seen in the families GP717B-33 and GP717B-189, in which numerous transgene-positive but herbicide-susceptible progeny were detected (Table 1).

Failure of transgenic gametes to compete with nontransgenic gametes is a potential explanation for the recovery of fewer than expected transgenic progeny. Zhang et al. showed that transmission of transgenes through male gametes was lower than through female gametes. It is possible also that the PCR assays failed to detect transgenic sequences in some plants; however, Southern hybridizations on DNA of selected plants using probes specific to bar showed perfect correlation to the PCR data. Furthermore, these tests were based not just on the detection of bar (which can be difficult to detect via PCR) but on detection of the BYDV CP gene. We have found PCR assays for the CP gene to be very reliable. Instances where PCR assays of resistant plants failed to detect transgenic sequences were rare.

Transgene Behaviour in Progeny

Two transgenic sublines from GP717B-4-9, which showed the expected patterns of transgene segregation, and one transgenic subline from GP717B-59-11, which was extremely resistant to glufosinate-ammonium damage based on visual assessments. Several sublines from each of these lines were used as the transgenic parents of single cross and backcross populations derived from crosses to a variety of elite cultivars and breeding lines used (Table 2). Crosses and backcrosses were made also to nontransgenic Golden Promise.

Segregation for herbicide resistance among progeny of these parents generally fit Mendelian expectations for a single, dominant locus (Table 3). The segregation for herbicide resistance among F_2 and BC_1F_1 progeny to be, respectively, 3:1 R:S and 1:1 R:S (note that transgenic parent of the BC_1F_1 progeny was the hemizygous F_1). The exception was the unexpectedly high numbers of resistant segregants among the F_2 progeny from the one cross involving GP724B-4-9 T_3 #37.

Table 2. Germplasm involved in crosses and backcrosses with transgenic parents

Transgenic parent subline	Nontransgenic parent (NTP)	Comments on NTP
GP717B-59-11 T_3 #19	'Garnet'(♂)[†]	Western U.S. 2-rowed malt cultivar
	'Baronesse' (♂)	German 2-rowed feed cultivar
	88Y394 (♂)	Western U.S. 6-rowed breeding line
GP724B-4-9 T_3 #30	'Golden Promise' (♂)	U.K. malting cultivar, mutant derived
	'Golden Promise' (♀)	Western U.S. 2-rowed
	90Ab321 (♂)	breeding line with CIMMYT-derived parentage
	'Garnet' (♂)	
GP724B-4-9 T_3 #31	90Ab321 (♂)	
	'Golden Promise' (♀)	
GP724B-4-9 T_3 #32	'Stander' (♂)	Midwestern 6-rowed malt cultivar
	'Stander' (♀)	
	'Garnet' (♂)	
	'Golden Promise' (♂)	
GP724B-4-9 T_3 #37	'Golden Promise' (♂)	

† Symbol (♂ or ♀) indicates that the nontransgenic parent was used as a female or male, respectively, for the first cross. The nontransgenic parent was used as female in all backcrosses.

To further examine segregation patterns for glufosinate-ammonium resistance, additional progeny tests were conducted on each of the cross- and backcross-derived populations. For each population, F_3 and BC_1F_2 progeny derived from randomly selected, resistant F_2 and BC_1F_1 plants (12 for each generation) were treated with 0.05% glufosinate-ammonium.

Table 3. Segregation for glufosinate-ammonium resistance in F_2 and BC_1F_1 progeny of crosses of transgenic barley with nontransgenic barley

Transgenic parent subline	No. of crosses	F_2 segregation	BC_1F_1 segregation
GP717B-59-11 T_3 #19	3	64R:20S	22R:23S
GP724B-4-9 T_3 #30	4	86R:26S	29R:26S
GP724B-4-9 T_3 #31	2	40R:16S	16R:12S
GP724B-4-9 T_3 #32	4	88R:25S	23R:24S
GP724B-4-9 T_3 #37	1	23R:4S*	16R:11S
all lines	14	301R:91S	106R:96S

* Segregation differs from expected ratios of 3:1 (F_3 segregation) and 1:1 (BC_1F_1 segregation) P = 0.05, as determined by chi-square analyses.

No differences in segregation patterns were noted for different populations derived from the same transgenic parent, and data from the various crosses to a particular parent were therefore pooled (Table 4).

Table 4. Segregation for glufosinate-ammonium resistance in progeny tests of resistant F_2 and BC_1F_2 plants.

Transgenic parent/ Generation	No. of heterozygotes: No. of homozygotes†	Segregation of heterozygous plants‡
GP717B-59-11 T_3 #19, F2	26:10	145R:59S
GP717B-59-11 T_3 #19, BC1F2	24:12	276R:94S
GP724B-4-9 T_3 #30, F2	39:9*	217R:83S
GP724B-4-9 T_3 #30, BC1F2	35:12	404R:152S
GP724B-4-9 T_3 #31, F2	23:7	72R:28S
GP724B-4-9 T_3 #31, BC1F2	18:6	210R:75S
GP724B-4-9 T_3 #32, F2	32:15	378R:89S*
GP724B-4-9 T_3 #32, BC1F2	29:21	620R:125S*

Contd...

Contd...

GP724B-4-9 T_3 #37, F2	7:5	35R:11S
GP724B-4-9 T_3 #37, BC1F2	8:4	98R:29S

* Segregation differs from expected ratios, P = 0.05, as determined by chi-square analyses.
†Selection of resistant plants only for testing results in the expectation of 2 heterozygotes:1 homozygote.
‡Expected segregation from heterozygous plants is 3:1 resistant:susceptible.

In general, expectations were met both for the number of homozygous and heterozygous individuals, and for the segregation patterns of heterozygous plants. The exceptions, again, were several instances of more resistant segregants than expected.

The very large deviations from expectations among progeny of crosses to subline GP724B-4-9 T_3 #32 are inconsistent with the normal patterns of inheritance observed among F_2 and BC_1F_1 progeny of this subline, and an adequate explanation for this observation remains elusive.

Herbicide Resistance

Data on the dry weight of seedlings, expressed as a percentage of the dry weight of untreated control plants. Symptoms of herbicide damage generally were evident within several days, and after 1 wk many of the nontransgenic controls sprayed with the 0.5 g kg-1 (0.05%) glufosinate-ammonium began to show signs of recovery and regrowth. Linear relationships did not exist for some of the treated groups (for instance, the transgenic parents did not exhibit reductions in dry weights except at high levels of herbicide treatment), so analysis of variance was used instead of linear regression.

The nontransgenic parents were clearly damaged by herbicide treatment, and at the highest treatment levels

(5.0 and 10.0 g kg^{-1} [0.5 and 1%] glufosinate-ammonium) most plants rapidly ceased growth, exhibited severe damage, and were dead or nearly dead 1 wk after treatment. No differences in herbicide resistance were observed among the nontransgenic breeding lines and cultivars; all appeared equally susceptible. The differential control responses at the 0.5 g kg^{-1} (0.05%) treatment level cannot be directly compared. Note that different cultivars comprise the nontransgenic controls in these two data sets, and these data were collected from experiments conducted at different times in the greenhouse. The transgenic parents exhibited obvious resistance to herbicide treatments and showed no statistically significant damage ($P = 0.05$) even at the highest treatment level, although there was a trend toward lower dry weights and increased damage as herbicide concentration increased.

Differences were observed in the relative level of herbicide resistance between progeny of the two transgenic lines. Transgenic progeny derived from crosses to GP717B-59-11 subline showed the highest levels of resistance, and were similar in their reactions to line GP717B-59-11; this is consistent with earlier, visual observations of a high level of resistance in this line. The progeny of crosses to the GP724B-4-9 sublines were relatively less resistant. Interestingly, and unexpectedly, the progeny of the two different sublines within GP724B-4-9 exhibited differences in herbicide resistance that were obvious from both visual observations and analyses of dry weights. No differences were detected among crosses to the same transgenic parent, nor between the two different populations (F_2 and BC_1F_1).

Progeny derived from crosses to subline GP724B-4-9 T3 #30 exhibited resistance very similar to that of subline GP724B-4-9 T3 #30, although there was a significant decrease in dry weight of progeny lines at the 10.0 g kg^{-1} treatment level not seen for the transgenic parent. Note, however, that the numerical differences between progeny and transgenic parent performance for this treatment are slight, and that the performance of subline GP724B-4-9 T3 #30 was based three data points (three reps) vs. 24 data points (three reps × four crosses × two generations) for the transgenic progeny. Additional replication of T_3 #30 likely would show statistically significant damage at the 10.0 g kg^{-1} treatment level.

Genetic Relationship of Stalk Strength

Stalk Lodging resistance is an important aspect of plant standability in maize. Stalk lodging is breakage at or below the ear, which may result in loss of the ear at harvest. Several methods have been devised to measure stalk strength as a means of improving stalk lodging resistance. Sibale et al. described use of a modified electronic rind penetrometer to measure stalk strength. Rind penetrometer resistance (RPR) was negatively correlated with stalk lodging. Recurrent S_0 plant selection for RPR has been effective in separating the synthetic population MoSCSSS into two distinct subpopulations. The rind penetrometer, therefore, is a valuable tool for measuring stalk strength.

Thompson found that lower internodes had greater stalk strength than higher internodes. Chesang-Chumo reported that ear height (EH) was decreased by 22.5% over five cycles of selection for high RPR in MoSCSSS. The correlation between RPR and EH is of interest in terms of direct vs. indirect selection for stalk strength. That is, to what extent does selection for high RPR increase stalk strength per se, or does decreasing EH result in increased stalk strength?

Little research has been conducted to investigate the nature of QTL for correlated traits. Paterson et al. found a chromosomal region affecting both mass per fruit and soluble solids in tomato (Lycopersicon esculentum Mill). This region contained the Self-pruning Protein (sp) locus which was shown to affect both traits. Schön et al. adjusted tunnel length (TL), a measure of European corn borer (ECB, Ostrinia nubilalis Hübner) damage, for plant height (PH). They found that four of the seven original TL QTL remained significant. One of the three QTL that lost significance after adjusting for PH overlapped a PH QTL. The authors determined that that particular QTL was not responsible for a physiological mechanism for resistance to ECB, but rather, was limiting TL since PH determines the total length of tunnels possible.

Several statistical programs have been developed to map QTL for multiple traits. In the multiple trait analysis method proposed by Jiang and Zeng, multiple traits are mapped simultaneously, taking advantage of the correlation between traits by considering the correlated trait data as repeated measurements of a single trait. In doing this, parameter estimation is improved and statistical power is increased. Because it was our goal to reduce correlated trait effects on QTL detection to obtain a truer genetic description of the trait of interest, RPR, we decided analyze correlated traits in a way nearly opposite of the method of Jiang and Zeng.

Population Development

Recurrent S_0 plant selection for increased and decreased RPR within the synthetic MoSCSSS yielded MoSCSSS(H24-high rind penetrometer [HRP])C10 and MoSCSSS(H25-low rind penetrometer [LRP])C11, respectively. Recurrent S0 plant selection for low stalk crushing strength (another method of measuring stalk strength) within MoSQB (Gerdes et al., 1993) yielded MoSQB(S10)C6. To initiate population development without delay, samples of each of MoSCSSS(H24-HRP)C10 and MoSQB(S10)C6 were grown in a greenhouse ground bed and evaluated for RPR. We selected the plant with the highest RPR in the MoSCSSS(H24-HRP)C10 sample and the plant with the lowest RPR in the MoSQB(S10)C6 sample. These plants were self pollinated, and the resulting progeny are hereafter referred to as MoSCSSS-High1 and MoSQB-Low, respectively. Because there was no overlap between the two original populations [MoSCSSS(H24-HRP)C10 and MoSQB(S10)C6], we believe that adequate parental materials were selected. MoSCSSS-High1 was self pollinated and the derived progeny are referred to as MoSCSSS-High2 and MoSCSSS-High3.

Population 1 was formed by crossing an individual MoSQB-Low plant (female) with an individual MoSCSSS-High1 plant. Population 2 was formed by crossing an individual MoSCSSS-High2 plant (female) with an individual plant from MoSCSSS(H25-LRP) (hereafter referred to as MoSCSSS-Low). Population 3 was formed by crossing Mo47 (female) with an individual MoSCSSS-High3 plant. Population 4 was formed by crossing B73 (female) with Mo47. For all populations, F1 plants were self pollinated yielding F2 individuals, which were self pollinated to produce F2:3 families. Populations 1, 2, 3, and 4 included 282, 291, 291,and 244 F2:3 families, respectively. The F2:3 families were sib pollinated to increase seed for phenotypic data collection.

Populations 1 and 2 were designed specifically to map stalk rind strength QTL since both parents were selected for high and low stalk strength phenotypes. Population 3 was designed to map QTL for RPR and resistance to second-generation ECB with Mo47 as the source of resistance.

Phenotypic and Genotypic Data Collection

Population 1 was planted as three 10x10 triple lattices at each of three locations: Agronomy Research Center in 1999, and Tipton and Agronomy Research Center in 2000. Populations 2 and 3 were planted as three 10x10 triple lattices per population at each of four locations: two locations in Puerto Rico in 1999, and Tipton and Agronomy Research Center in 2000. Data collected included RPR, PH, and EH. Rind penetrometer resistance was determined for 10 competitive plants plot-1 with a modified electronic rind penetrometer.

About two weeks after flowering, plants were probed in the middle of the flat side of the internode below the primary ear node. Ear and plant heights were obtained by measuring the distance from the ground to the primary ear node and the collar of the flag leaf, respectively, for 10 competitive plants plot-1. Populations 1, 2, and 3 were genotyped with 89, 77, and 86 SSR markers, respectively, according to the procedures described in Flint-Garcia et al.

Data Analysis

Year-location combinations were treated as independent environments. Rind penetrometer resistance was adjusted for EH at each environment generating a new trait, RadjE, as follows: $RadjE_i = RPR_i - b_1 (EH_i - \overline{EH})$ where b_1 is the linear regression coefficient between RPR and EH. Environment means and broad-sense heritability estimates were calculated. Family means across environments were used to compute phenotypic correlation coefficients with SAS PROC CORR.

MAPMAKER/EXP version 3.0b was used to construct linkage maps. QTL Cartographer version 1.14d was employed for QTL analysis of family means across environments as described in Flint-Garcia et al. (this issue). For each trait-population combination, experiment-wise log-odds ratio (LOD) thresholds at P = 0.05 were determined by analyzing 1000 permutations of the data. A one-LOD drop from the peak position was used as a confidence interval for QTL location. Overlapping confidence intervals were used to determine QTL in common among traits.

The statistical program EPISTACY was used to test for the presence of epistatic interactions between marker pairs at P < 0.001. To build multilocus models, markers nearest to single-effect QTL and those involved in epistatic interactions were subjected to stepwise regression at P < 0.05 by SAS PROC REG. Markers were added to the model in order of increasing significance (forward regression in), and were removed if their significance while in the model exceeded 0.05 (backward regression out).

Developing Perennial Upland Rice II

In South Asia, annual crops of upland (dryland) rice (O. sativa) are typically grown in hilly areas, leading to problems associated with soil erosion and watershed degradation. Perennial cultivars of upland rice, if developed, could help subsistence farmers improve food security and protect the environment. Perennial upland rice would reduce erosion of soil from hilly farms, and protect downstream reservoirs from excessive silting by providing long-term ground cover. These advantages would be realized while providing a preferred source of food.

Though cultivars of Asian (O. sativa) rice vary in ratooning ability when grown in paddies, the species is only weakly perennial, especially under upland conditions. Thus, genes from strongly perennial wild relatives of O. sativa will be needed to realize our goal of developing a perennial cultivar for the uplands.

Oryza longistaminata (2n = 24, AA) is a potentially useful source of genes for developing perennial upland rice. Oryza longistaminata is a highly diverse, allogamous species that is broadly distributed throughout Africa. Typically, accessions of O. longistaminata are strongly perennial and have vigorous rhizomes. However, natural populations of O. longistaminata mostly grow in standing water, so they may lack adaptation to upland conditions. Oryza sativa and O. longistaminata are considerably diverged as compared with other AA species. To obtain F1 interspecific progeny and the first few backcross generations, in vitro embryo rescue is usually required.

Briefly, a single F1 hybrid between the indica land race, BS125, and the O. longistaminata genotype, WL02-15, was used. Because the F_1 did not produce fertile pollen, it was used as a female parent in crosses with 11 O. sativa cultivars (Azucena, Basmtai370, Cuiabana, IR60080-46A, IR63371-38, IR63380-08, IRAT104, IRAT212, IRAT216, Vandana, WAB56-50) to produce BCS progeny. The F_1 was also crossed with five O. longistaminata genotypes (DL01-1, SL313-13, SL313-19, WL02-2, and WL02-15), to produce BCL progeny. Alain Ghesquière kindly provided the F_1 and O. longistaminata genotypes. The O. sativa parents and the BCS progeny lacked rhizomes. The O. longistaminata parents and the BCL progeny were highly rhizomatous. The BCL progeny were selfed to produce BCL F_2 progeny.

BCL F_2s that had rhizomes and acceptable pollen fertility were crossed as males with BCS F_1s to give the first intermated generation. The progenies were obtained from the first through third cycle of intermating. Genotypes that were identified as rhizomatous, on the basis of the distance between shoots at the soil surface, were preferentially used as parents for each cycle of intermating. Plants obtained from additional backcrossing to O. sativa were also included in the intermated population. On the basis of their pedigrees, the progenies were expected to have 50 to 75% O. sativa alleles. S_1 seed was obtained from bagged panicles. Oryza sativa cultivars were chosen primarily for adaptation to upland production. With the exception of UPL-Ri-5, which was included as a high-yielding check, the cultivars evaluated were parents in the intermated population.

Height Trials

Two entry x irrigation area trials and one postharvest cutting height trial were conducted. To comply with Philippine plant quarantine regulations, the O. longistaminata parental genotypes were not included in field trials. Seeds for all trials were sown in flats on 29 July 1999 and grown in a glasshouse. Seedlings were transplanted to the field on 14 or 16 Aug. 1999. Plots were 3 m long. Rows were spaced 30 cm apart and within each row, plants were spaced 25 cm apart, for a total of 12 seedlings. The trials were planted on a gently sloping upland field at IRRI's headquarters (14°11' N, 121°15' E), Los Baños, Laguna Province, Philippines.

The irrigation areas were opposite sides of a 1-ha field. During the dry season (January through May), the field was separated into a wet side and a dry side by an 18-m-wide strip of nonirrigated land. The wet side was sprinkler irrigated for approximately 5 h once per week. The soil on the dry side was saturated by 7 to 16 h of sprinkler irrigation over 1 to 2 d starting on 29 Feb., 30 March, and 2 May 2000. During the rainy season (June through December), supplemental irrigation was not needed. Thus, the main difference between irrigation areas was availability of water during the dry season.

Entry × Irrigation Area Trials

The two entry x irrigation area trials were lattice (0,1) designs with two irrigation areas, two or three replications per irrigation area (depending on the trial), and 10 blocks per replication. Plots were single rows. The trial with two replications compared 14 check cultivars and 51 S_1 families. The trial with three replications compared 16 check cultivars and 54 S_1 families. Sixteen S_1 families were common to both trials. Availability of seed determined the trial in which the entries were used. Upon harvesting, plants were cut 25 to 30 cm above the ground and allowed to regrow.

Cutting Height Trial

The effect of cutting height at harvest on postharvest survival. The cutting height trial was a split plot design, with three postharvest cutting treatments assigned to whole plots, and 14 entries assigned to subplots. There were three replications of whole plots. Subplots were 2 rows containing 24 plants. All plots in the cutting height trial were on the dry side of the field. Cutting treatments were high = stems uncut with only panicles removed, medium = stems cut half way between the soil surface and the flag leaf, and low = stems cut 15 cm above the soil surface. Plants were cut in December 1999 after the initial harvest and again in March 2000 after harvest of the ratoon crop. Entries consisted of two cultivars and 10 S_1 families. Five S_1 families were common to the cutting height trial and the entry x irrigation area trials.

Data on vegetative vigor, survival and presence of rhizomes were recorded. Individual plants in plots were measured. IRRI's 1 to 9 scale for vegetative vigor was modified to include a value for dead plants: 1 = extra vigorous, 3 = vigorous, 5 = normal, 7 = weak, 9 = very weak, 10 = dead. Initial vigor was recorded in October 1999, before the first harvest. Approximately 1 yr after planting, vigor and survival were recorded on 2 and 3 Aug. 2000 for the entry x irrigation area trials, and on 15 Aug. 2000 for the cutting height trial. For the cutting height trial, vigor and survival also were recorded approximately seven months after planting on 23 March 2000. Presence of rhizomes was recorded for all trials through the first 6 mo after transplanting. Plants that had at the soil surface a distance between consecutive shoots of 3.5 cm or more were scored as rhizomatous.

To avoid loss of seed from shattering and to comply with Philippine plant quarantine regulations, panicles of interspecific progeny were enclosed in nylon net bags (approximately 15 by 30 cm). Panicles of cultivar controls were also enclosed in net bags but some bird damage was observed before bagging was completed. Panicles were harvested when mature. For entry x irrigation area trials only, data on yield, awning, and threshability were also taken. Harvested seed from each plant was weighed. Plant height was recorded between 11 and 14 Oct. 1999. The lengths of three awns from each plant were measured. Panicle threshability was determined after harvest and drying by means of IRRI's standard, 1 to 9, scale; 1 = difficult (<1% shattering), 9 = easy (51–100% shattering).

Analyses of variance were conducted with SAS procedure MIXED, using type III sums of squares. For the entry x irrigation area trials, entry type (cultivar or progeny), entries, and irrigation area (wet or dry side) were considered fixed effects; replications nested in irrigation areas and blocks nested in replications were considered random effects. For the cutting height trial, cutting treatment and entry were considered fixed effects; replications were considered random effects. For all analyses, interactions between fixed and random effects were considered random, degrees of freedom were calculated via the Satterthwaite option, and means were obtained with the lsmeans statement.

Seed-Oil Concentration in Sunflower

Seed-oil concentration (achene-oil concentration) is a major consideration in sunflower breeding because it is one of the two components of oil yield. The genetic basis of seed-oil concentration has been described to a limited degree. Seed-oil concentration has relatively high broad sense heritability (0.6–0.7) and high narrow sense heritability (0.5–0.6) on a single plant basis. Some research has been done to study gene action involved in the expression of the traits. Highly significant additive, dominant, and epistatic effects were determined for seed-oil concentration in two F_2 populations and their reciprocal backcrosses. Estimates of general (GCA) and specific combining ability (SCA) indicated additive effects were more important for seed-oil concentration.

Significant additive effects were also reported for seed oil and seed percentages, although significant dominant effects for seed percentage were detected in irrigated conditions. Additive gene action has also been reported for hull percentage. In general, these results indicate that additive effects predominantly influence oil concentration in the whole seed and its components. However, some non-additive effects also seem to affect these traits. The high heritability and predominately additive gene action of this trait facilitate selection in early generations of inbreeding and cultivar development.

Improvement of seed-oil concentration in hybrid progeny has been achieved by increasing seed-oil concentration in the inbred progeny. This improvement has been accomplished by a reduction of the hull percentage (or increase of seed percentage), and to a lesser extent, by an increase of the seed oil concentration. The use of molecular markers provided additional information about the genetic basis of seed-oil concentration. Leon et al. used a genetic map of 201 RFLP loci and oil data collected from individual plants in the F_2 generation to locate six QTL associated with 57% of the genetic variation. Two of the QTL were related to seed oil concentration (linkage groups C and I), two to seed percentage (linkage groups G and J), and the other two to both traits (linkage groups B and N). Additive gene action was predominant for seed-oil concentration and its components.

Leon et al. reported a dominant factor (Hyp) determining the presence of white pigments in the hypodermis that was located in linkage group 'G'. Seeds with white hypodermis had lower oil concentration than those with unpigmented hypodermis. The Hyp factor was located in the same map interval as one QTL with major effects on seed-oil concentration. Mestries et al. also mapped QTL affecting seed oil content and found two to three QTL for this trait, depending on the environment and generation. These QTLs were associated with 19 to 54% of the phenotypic variance across generations.

Seed-oil concentration of sunflower is sensitive to environmental conditions during the grain-filling period. For example, reductions in oil yield and seed-oil concentration may occur when low temperatures and radiation prevail during the seed-filling period. At high latitudes, such conditions are frequently encountered by late-flowering genotypes and could be a source of genotype x environment interaction for seed-oil concentration.

Therefore, estimates of QTL positions and effects for seed-oil concentration in sunflower may be enhanced through analyses of phenotypic data collected from replicated progeny evaluated in suitable target environments, and through the coincident evaluation of traits that could alter our assessment of allelic effects on the primary trait.

Germplasm and Trait Characterization

The nonrestorer (B) lines ZENB8 (female) and HA89 (male) were crossed to produce the progeny. The F2 seed was produced by self-pollinating a single F_1 plant. ZENB8, a proprietary inbred line, has a seed-oil concentration of approximately 330 g kg^{-1} and reaches anthesis (flowers) approximately 75 d after planting at the photoperiod (15–16 h) and temperatures typical of the growing seasons at locations used. HA89, an inbred line released by the USDA, has a seed-oil concentration of 490 g kg-1 and flowers approximately 65 d after planting under the same conditions at those locations.

The F2-generation seed was hand planted in rows at Fargo on 14 May 1992, at a rate of two seeds per hill. Seedlings were removed to leave one plant per hill. Hills were 0.30 m apart within a row. The rows were 6 m long and 0.75 m apart. Five rows of each parent and the F_1 were planted at different periods (-10, -5, 0, +5, +12 d relative to the planting date of the F_2 seed) to estimate the within-row error variance. Before anthesis, individual heads of 235 F_2 plants were covered with pollination bags to ensure self-pollination and production of F_3-generation seed.

Single-row plots of the 235 F_3 families were hand planted at Daireaux, Venado Tuerto, and Balcarce on 17, 18, and 20 Nov. 1992, respectively. Fifteen plots of each parent and the F_1 generation were included at each Argentine environment to provide an estimate of the error variance within and across locations. Plots were 3 m long and contained 10 hills. The space between plots was 0.70 m. Three seeds per hill were planted and seedlings were removed to leave one plant per hill. The families, parents, and F_1 genotype were randomly assigned to plots at each location.

Seed-oil concentration (defined as oil weight/seed weight) was measured with a Nuclear Magnetic Resonance (NMR) analyzer. Measurements were made on a dried sample of 10 g of F_3 seed of each F_2 plant as described in Leon et al.. In the F_3 families, oil data were collected in the same way but from a balanced bulk of F_4 seed harvested from the F_3 plants in each row. The balanced bulk was created by taking equal volumes of seed from each plant in the plot.

Sunflower growth stages are defined according to Schneiter and Miller. Days from emergence (VE) to flowering, or anthesis (R5.5; when 50% of the flowers of a capitula are open) were recorded for F_2 plants and their corresponding F3 families at each environment, as described in Leon et al. The day of flowering (DTF) of an F_3 family was considered as the day on which 50% of the plants reached the R5.5 stage. The DTF was used to quantify a portion of the life cycle of the parents, F_1 and F_2 generations, and F_3 families in these environments. The mean daily temperature during the period between flowering and physiological maturity was obtained from the nearest meteorological station. The period began when the first progeny reached anthesis and ended 45 d after the date of the anthesis of the latest progeny.

The 205 RFLP loci covered 1380 centimorgans (cM) and were arranged in 17 linkage groups, the haploid number of chromosomes in this species. The average interval size was 5.9 cM. The genetic map was constructed using MAPMAKER version 3.0. Genotypic classes at 23 loci deviated significantly from the expected ratios. Those loci exhibited a deficiency in the ZENB8 homozygous class. The majority of the loci with deviant ratios (18/23) were located to four regions, representing linkage groups G, L, and P.

Simple-Pearson phenotypic correlations between seed-oil concentration and DTF were calculated for each location and for the average values for all locations (i.e., the mean environment). Broad-sense heritability was estimated according to Allard for individual plants in the F_2 generation. The within-row variance in the F_2 generation was estimated by pooling within-row variances of the parents and F_1 genotype. The error variance among rows was estimated in the F_2 generation. Genetic variation was then estimated by subtracting the within- and among-row variances from the phenotypic variance.

For the F_3 families, broad-sense heritabilities were estimated using variance components according to Fehr; for the heritability on plot basis (for each location) $h^2 = \frac{\sigma^2_g}{\sigma^2_e + \sigma^2_g}$ and heritability on entry- mean basis (across locations) $h^2 = \frac{\sigma^2}{\sigma^2_e/rt + \sigma^2_{gxe}/t + \sigma^2_g}$, where t and r are the number of environments and replications within environments, σ^2_e the experimental error variance, σ^2_g the genotypic variance, and σ^2_{gxe} the genotype x environment interaction variance. Estimates of σ^2_e within and across locations were obtained from the parents and F_1, the genotype x environment (GxE) interaction was tested according to Hallauer and Miranda by means of the 2e estimated from the parents and F_1 genotype across locations.

Composite interval mapping (CIM) was used for mapping QTL. Phenotypic data consisted of the seed-oil concentration for each F_2 plant or F_3 family evaluated at each location and the average value of each F_3 family across locations (herein, the mean environment). The use of single replicates of each family in multiple environments has been described previously for QTL mapping maize for grain yield and plant height and in sunflower for days to flowering and photoperiod response.

Initially, an analysis was made with the first statement to check the database for errors and outliers. A second analysis with the model D and scan statements with a LOD threshold value of 2.5 was conducted to select cofactors. A third analysis was done adding the preselected cofactors in the cov statement and the smodel statement for detection of digenic epistatic interactions between QTL with significant main effects. The coefficient of determination (R_2) from the model for the mean environment was compared with the estimated broad-sense heritability to estimate the amount of genetic variation associated with RFLP loci in multiple regression.

Epistatic effects among all pairs of loci were assessed with two-way Analyses of Variance using the program EPISTACY. A total of 20 910 pairwise tests were performed and the interlocus interaction variance was partitioned into additive x additive, additive x dominance, dominance x additive, and dominance x dominance interactions. Because of the high number of comparisons, an epistatic interaction was declared when the F-test was significant at the P0<0.0001 level. Interactions were then added to the multiple regression model in the seq statement of the PLABQTL program to estimate the amount of genetic variation associated with the complete model.

Estimates of the additive (a) and dominant (d) effects were obtained by fitting a model including all QTL as described in Bohn et al.. The d/a (dominant/additive) ratio scale described by Edwards et al. was used to classify gene action [A = additive or partial dominance ($0 < |d/a| < 0.55$); D = partial dominance or dominance ($0.55 < |d/a| < 1.20$), OD = overdominance ($|d/a| > 1.20$)].

6

Oilseeds

New Industrial Oilseed Crops

Developing new oilseeds is several-fold. First, a new crop can provide a significant increase in the level of income for the farmer; second, new crops can contribute to the alleviation of negative balance of payments; and third, they can provide materials that are critical to our national interests, especially if we at present depend on imported materials. The reason we select specific plant species for development is the difference in chemical functionality of the proposed new seed oil from the oils we now have domestically available.

Medium Chain Fatty Acids—Cuphea

The basic source of medium chain fatty acids is tropically-grown coconut and palm kernel oil. The U.S. has no domestic source of these materials and therefore imports about 500,000 tonnes (one billion pounds) annually of "lauric" oils. However, many species from the genus *Cuphea* (Lythraceae) have potential as sources of medium chain triglycerides. These plants are native to the New World, from Southern U.S. to Northern South America. Most are herbaceous annuals that will grow in many locations. However, *Cuphea* is only a few years from the wild and still has the characteristics of a wild plant. Those characteristics that differ from cultivated plants are its propensity to seed shatter, its indeterminate flowering nature, and its overall stickiness. If these wild traits can be overcome, *Cuphea's* chemistry, coupled with the annual and therefore renewable nature of the plant, certainly can make it a new crop.

Table 1 illustrates the diversity in fatty acid composition available in Cuphea germplasm. While there is some variation from accession to accession, the table shows species that are rich in specific single fatty acids. Cuphea painteri, for instance, is very rich in caprylic (8:0) acid (73%) while C. carthagenensis has lauric acid (12:0), as its major fatty acid (81%).

Table 1. Fatty acid composition of some Cuphea seed oils.

	Distribution (% of total fatty acids)				
Species	8:0 caprylic	10:0 capric	12:0 lauric	14:0 myristic	Others
C. painteri	73.0	20.4	0.2	0.3	6.1
C. hookeriana	65.1	23.7	0.1	0.2	10.9
C. koehneana	0.2	95.3	1.0	0.3	3.2
C. lanceolata		87.5	2.1	1.4	9.0
C. viscosissima	9.1	75.5	3.0	1.3	11.1
C. carthagenensis		5.3	81.4	4.7	8.6
C. laminuligera		17.1	62.6	9.5	10.8
C. wrightii		29.4	53.9	5.1	11.6
C. lutea	0.4	29.4	37.7	11.1	21.4
C. epilobiifolia		0.3	19.6	67.9	12.2
C. stigulosa	0.9	18.3	13.8	45.2	21.8
Coconut	8	7	48	18	19

Cuphea koehneana is probably the best example of a monoacid seed oil, with more than 95% of its acyl groups as capric acid. As a source for lauric acid, Cuphea ssp. have more to offer than coconut oil (Table 1), because the concentration of lauric acid in the oil is potentially much greater. Isolation of single fatty acids should be easily accomplished and tailor-made fatty acid compositions should be possible.

Oil and protein values were determined only on a handful of accessions because only a few seeds have been collected from many of the species. In those species, oil percent varied from 16 to 42%. Protein (%N x 6.25) levels were from 15 to 24% of the whole seed. Table 2 shows the amino acid composition of two *Cuphea* species.

Table 2. Amino acid compositions of some potential oil seed species.

	Content (g/16 g of protein)			
Amino acids	*Cuphea painteri*	*Limnanthes douglasii*	*Lesquerella fendleri*	*Simmondsia chinensis*
Aspartic acid	8.0	8.0	7.2	10.0
Threonine	3.0	4.3	4.5	6.0
Serine	4.9	4.4	4.6	6.9
Glutamic acid	15.3	16.3	13.7	10.4
Proline	3.5	4.2	6.7	6.0
Glycine	4.5	6.1	5.9	16.2
Alanine	3.9	4.4	4.5	6.3
Valine	4.6	5.0	4.8	5.4
Cystine	1.2	1.4	1.8	3.5
Methionine	1.7	1.4	1.3	0.9
Isoleucine	3.9	3.8	3.6	3.1
Leucine	6.2	6.9	5.8	6.8
Tyrosine	2.7	2.2	3.0	3.6
Phenylalanine	3.8	3.7	3.8	3.3
Lysine	3.9	6.9	6.6	4.3
Histidine	2.2	2.2	2.5	1.7
Arginine	10.2	7.5	7.9	5.6

The triglyceride analysis of *C. lanceolata* shows that the combination of acyl groups within triglyceride molecules is not the result of random distribution but the specific combination of certain fatty acids, particularly C10. Gas chromatography of the intact triglycerides show that the C8 was completely contained in the group with carbon number of C28 indicating that the C30 peak was made up of three C10 fatty acids and not a combination of C8, C10, and C12.

These combinations have significance when *Cuphea* oils are to be used in nutritional and medical applications. Other members of the Lythraceae have been examined to see if fatty acid compositions are similar to *Cuphea*. However, only small amounts of medium chain fatty acids were found.

Seed Oils from Apiaceae (Umbelliferae)

While the seeds oils from the plants of the Apiaceae (Umbelliferae) do not contain medium chain fatty acids per se, most do contain large amounts of petroselinic (*cis*-6-octadecenoic) acid. This positional isomer of the common oleic (*cis*-9-octadecenoic) acid can be cleaved at the double bond to form two industrially useful materials, adipic and lauric acids. The method of choice to perform this type of cleavage is ozonolysis. For example, the production of pelargonic and azeleic acids is from the ozonolysis of oleic acid. However, the production of lauric acid by this means will have to compete with lauric acid from a much cheaper starting material, coconut oil.

With the possible exception of dill, most Umbelliferae are not produced in the U.S. for their seed oil. Production of such crops as celery, dill, carrots is for food and or condiments. However, seed yields for common fennel of over 2800 kg of seed per hectare have been reported. With agronomic and breeding research, higher seed yields and oil content should be achieved. Wild species of the Apiaceae have been reported that have as much as 40% seed oil containing 80% petroselenic acid.

Meadowfoam **(Limnanthes alba)**

The oil composition from meadowfoam seed is unique in several ways. First, over 95% of its acyl groups are longer than C18; secondly, about 90% of these fatty acids have double bonds in the delta-5 position; and thirdly, there are essentially no polyunsaturates. Along with the 60% *cis*-5-eicosenoic acid, the 8% *cis*-5-docosenoic acid, and the 10% erucic acid one finds 15-30% of 22:25,13. The latter is a polyunsaturated dienoic acid, but six carbon atoms separate the double bonds, thus making the acid react essentially as a monoenoic fatty acid in terms of its oxidative stability.

The oil should be oxidatively stable, not only because of the lack of polyenoic fatty acids and its long chain nature but also because the delta-5 bond is more stable than olefins with the double bond in the center of the fatty acid molecule. The *cis*-5 bond also allows lactonization to proceed easily, especially after epoxidation. Oxidative cleavage of the 20:15 at the double bond results is glutaric and pentadecanoic acids. The use of the latter 15:0 acid should be evaluated because it is not now industrially available. The 22:25,13 has been successfully epoxidized. The reactive diepoxide should be a useful industrial intermediate.

Natural meadowfoam oil is in the form of triglycerides. However, they have been converted to liquid wax-esters by reduction of fatty acids to alcohols and reesterification with the unreacted fatty acids. These C40 and C42 molecules should be useful, both sulfurized and unreacted, in lubricants. Meadowfoam (*Limnanthes alba*) is a member of the Limnanthaceae. Though the presence of the sulfur containing glucosinolates is usually associated with the Brassicaceae (Cruciferae), meadowfoam seed contains significant levels of glucosinolates that upon hydrolysis yield both a volatile mustard oil, *m*-methoxybenzyl isothiocyanate, and the nonvolatile 5,5-dimethyloxazolidine-2-thione. These products must be considered when meadowfoam defatted meal is used as animal feed. Meadowfoam is grown as a winter annual principally in the Willamette Valley of Oregon where yields of over 1.1 t/ha are routine.

Lesquerella

Plants of the genus *Lesquerella* are members of the Brassicaceae (Cruciferae) and are native to North America. Their seed oils are rich in hydroxy fatty acids. Castor oil is the only commercial source of hydroxy acids, and the fatty acid in the oil is ricinoleic, 12-hydroxy *cis*-9-octadecenoic, acid, found at the 85% level. Though most *Lesquerella* species have small amounts of this acid, large concentrations of other hydroxy acids are found. Though most species in this genus have several hydroxy fatty acids, they are generally rich in only one.

The dienoic hydroxy fatty acids, densipolic and auricolic, have potential as fatty intermediates and tung oil replacement. At the USDA's Water Conservation Laboratory, Phoenix, Arizona, agronomic and breeding research is now underway on one of these, *Lesquerella fendleri*. Yields of better than 1500 kg/ha have been achieved in just a few years. With continued work in this area increases in seed yields, lesquerolic acid, and oil content are expected.

The structure of lesquerolic acid is homologous to ricinoleic acid. It has two additional carbon atoms at the carboxyl end of the molecule. This similarity allows many of the same reactions and, presumably, the same uses as for the castor-based fatty acid. This includes the production of 2-octanol and 2-octanone, depending on reaction temperature during alkali cleavage of the acid. However, this same reaction produces dodecanedioic and hydroxydodecanoic acids from lesquerolic acid instead of the analogous acids from ricinoleic that are two carbons shorter.

Table 3. Seed characteristics of Limnanthes alba.

Seed characteristic	Content
Oil (%)	17-29
$20:1^{5}$ (%)	50-65
$22:1^{5}$ + $22:1^{13}$ (%)	10-29
$22:2^{5,13}$ (%)	15-30
Protein (%)	11-28
Glucosinolates (%)	3-10
Seed wt (g/1000)	4.2-9.8

The C12 dibasic acid, dodecanedioic acid, is the basis for nylon 1212. The current raw material for production of dodecanedioic acid is petroleum. Without alkali, a major product from thermal fragmentation of ricinoleic acid is 10-undecenoic acid. This material, after bromination and subsequent amination, is the starting material for nylon-11. The analogous 12-tridecenoic acid is produced from lesquerolic acid and could be used in the production of the monomer for nylon-13. The saturated, via hydrogenation, hydroxy acids, as their lithium soaps, could be useful in the production of greases. The intact triglycerides from *Lesquerella* should be suitable for many uses. For example, they could substitute for dehydrated and oxidized castor oil. Many of the properties may be enhanced over those of castor oil because of the increased chainlength of this new crop oil.

Like most of its cruciferous relatives, *Lesquerella's* seed meal contains glucosinolates. There are several different glucosinolates identified in *Lesquerella* spp. Many of these can be removed easily from the meal if converted to the volatile isothiocyanate. The amino acid composition of the seed meal has been reported. Its especially high lysine content makes it attractive as a protein supplement.

*Jojoba (*Simmondsia chinensis*)*

Made up almost exclusively of liquid wax-esters, the oil of jojoba seed is unique in the plant kingdom. Jojoba is often touted as a whale oil substitute, while in fact it has features that, in most applications, make it superior. First of all, whale oil has a significant (about 30%) amount of polyunsaturated triglycerides as part of its total lipids. These make it oxidatively less stable than jojoba, which has essentially no triglycerides, polyunsaturated or otherwise. Secondly, jojoba wax-esters are considerably longer in chainlength, ranging from C40 to C44. The bulk of sperm whale oil has chainlengths of C32 to C36. The fatty acid and alcohol compositions, making up the wax-esters of jojoba oil, are listed in Table 4. The specific combinations of alcohols and acids have been established by mass spectrometry.

Table 4. Typical jojoba oil composition

Chainlength	Fatty acids (%)	Alcohols (%)
16	2	1
18	13	1
20	71	51
22	13	42
24	1	5

Jojoba oil's current use is centered on the cosmetic and personal care industry. This is due primarily to its ability to lubricate without the sense of greasiness. There are about 6,000 productive hectares (15,000 acres) of jojoba in the southwest U.S. As these areas become more fruitful, and additional areas come on-line, the price of the oil is expected to decrease. As this comes about, more industrially oriented uses for this material will emerge in the market place. A number of derivatives have been made. These include sulfurized and halogenated jojoba oil, for high-pressure applications such as in automobile transmissions, and hydrogenated jojoba as a wax.

The whole jojoba seed contains about 15% protein. When considering the large amount of oil in the seed (50%), the defatted meal has about 30% protein. However, the seed contains also about 11% antinutritional compounds. These are simmondsin, simmondsin 2'-ferulate, 5-desmethylsimmondsin, and didesmethylsimmondsin. Several methods are now under development to eliminate these materials in order to use the meal as a nutritional animal feed.

High-Erucic Acid Sources

Crambe *(Crambe abyssinica)* and industrial rape (Ackman 1983) (either *Brassica napus* or *B. campestris)* are potential sources for erucic acid in the U.S. While European rape has long been imported into this country, it is recently being introduced as a new crop for the U.S. Crambe has been suggested for many years as an erucic acid source and has been grown intermittently on small areas. All three species are members of the Brassicaceae (Cruciferae).

Erucic (*cis*-13-docosenoic) acid is now converted to erucamide for use as a slip agent in the manufacture of polyethylene sheets. The acid also has been cleaved to form brassylic and pelargonic acids. The brassylic acid has been used to produce nylon 1313. Both erucic acid itself and its cleavage product, brassylic acid, have potential in other commercial products such as plasticizers, lubricants, and surfactants.

The seed meals of these crops have high protein levels and good amino acid compositions (Table 3). While the glucosinolates in rapeseed can be eliminated through breeding, no germplasm is yet available to lower the 8% glucosinolates in crambe seed. However, crambe meal has been cleared for use at the 5% level in animal feeds.

Epoxy Oils

Many plants produce seed oils with epoxy fatty acids. A few produce as much as 80% of one epoxy acid, vernolic (12,13-epoxy-*cis*-9-octadecenoic) acid. Those with potential as crops are *Vernonia galamensis*, Stokes Aster, and *Euphorbia lagascae*. The fatty acid compositions of the seed oils of these species are given in Table 5.

Table 5. Fatty acid composition of epoxy oils.

	Content (% of total fatty acids)		
Fatty acids	*Vernonia galamensis*	*Stokesia laevis*	*Euphorbia lagascae*
16:0 (palmitic)	2.8	2.6	4.1
18:0 (stearic)	2.6	1.1	2.0
18:1 (oleic)	4.0	6.9	20.7
18:2 (linoleic)	12.8	13.7	9.2
18:3 (linolenic)	0.1	0.2	0.4
Vernolic acid	76.4	74.1	61.8
Others	1.3	1.4	1.8

None of these species are now grown in the U.S. However, *Vernonia galamensis* is grown experimentally in its native equatorial Africa. Until recently, the long daylight hours of the U.S. did not allow flowering of this species. Stokes aster (*Stokesia laevis*), a southeast U.S. herbaceous perennial, has been grown experimentally with promising results. *Euphorbia lagascae* is naive to southern Europe and is being considered as a crop there.

These epoxy oils have an advantage over commercially epoxidized oils in that the location, number, and configuration of epoxy and olefinic groups are rigorously known. The oil of *Vernonia galamensis* forms excellent baked coatings on steel and interpenetrating polymer networks with other polymers. These oils have potential in the production of plastics, paints, and lubricants.

Temperate Oilseed Crops

The United States imports several important industrial chemical feedstocks or derives them from petrochemicals or unspecialized seed oils, e.g., soybean oil. Notable examples are medium-chain, hydroxy, epoxy, and long-chain fatty acids. C12 fatty alcohols, for example, are derived from petrochemicals or imported natural oils. There are temperate oilseed crop sources of these fatty acids but few are commercially exploited. The past and present domestication activities in several unexploited temperate industrial oilseed crop species: *Cuphea* species, *Anethum graveolens* L. (dill), *Coriandrum sativuum* L. (coriander), *Foeniculum vulgare* Mill. (fennel), *Limnanthes alba* Bentham (meadowfoam), *Lunaria annua* L. (honesty or money plant), and *Stokesia laevis* Hill (Greene) (Stokes aster).

The fatty acids produced by these species include medium-chain (*Cuphea* ssp.), petroselenic (dill, coriander, fennel, and other Apiaceae), longchain (meadowfoam and money plant), and epoxy (Stokes aster) fatty acids. The industrial uses and chemistry, of these fatty acids have been reviewed. *Lesquerella fendleri* L., a source of hydroxy fatty acids, is a promising new industrial oilseed crop adapted to the desert Southwest (Thompson 1988). The status of *Lesquerella* domestication and commercialization is reviewed by Thompson in this volume. Yield data on *L. fendleri* and other species are needed to establish their potential for temperate agriculture.

New industrial oilseeds crops have, as a general rule, no domestication history beyond the last one or two decades. The lack of domestication histories is not surprising since their industrial use potential was discovered only within the last few decades. The rate of domestication of these species depends on the current agronomic characteristic of the crop. New oilseed crops generally have domestication barrier traits, e.g., seed shattering and seed dormancy. In addition, their germplasm resources may be limited. The rate of commercialization, ignoring factors unrelated to plant improvement, depends on the genetic variation for economically important traits and size of the germplasm resource. There are tremendous differences in genetic variation for traits affecting rates of domestication and commercialization among new oilseeds. These differences ultimately determine whether or not these species will be commercialized.

Cuphea

Lauric and capric acid (medium-chain fatty acids or MCFAs) have several industrial, medical, and nutritional uses. The main industrial use of lauric acid is in soaps and detergents. Lauric acid is of greater economic importance than capric acid. The United States relies on imports of seed oils rich in medium-chain triglycerides (MCTs) because there are no temperate lauric or capric oil crops. U.S. imports are comprised of coconut (*Cocos nucifera* L.), African oil palm (*Elaeis guineensis* Jacq.), and other tropical palm oils. The U.S. imported an average of 525,000 t/yr (1.16 billion lb/yr) between 1976 and 1983. World production of these oils is about 3,600,000 t/yr. MCT oils obviously, are of enormous global economic significance.

Cuphea is a genus of approximately 250 undomesticated species native to Mexico and Central and South America. One species, C. viscosissima, is native to the U.S. Several are adapted to temperate agriculture and have seed oils rich in capric and lauric acid. Thus, one possibility for domestic production of lauric and capric oils is through the cultivation of Cuphea. The economical cultivation of Cuphea is, however, presently prevented by the problem of seed shattering.

Seed shattering and seed dormancy usually determine the difference between undomesticated and domesticated forms of a species. The wild progenitors of several seed crops, e.g., wheat (*Triticum aestivuum* L.), rice (*Oryza sativa* L.), oats (*Avena sativa* L.), barley (*Hordeum vulgare* L.), and sorghum [*Sorghum bicolor* L. (Moench.)], are characterized by seed shattering. Nonshattering phenotypes are simply, inherited in these species and were presumably fortuitously discovered, selected, and fixed in populations sometime within the last 10,000 years. There is no apparent advantage to seed retention in natural populations; however, genetic variation for seed retention was obviously present in wild populations of these species.

Wild populations of *Cuphea* are characterized by seed shattering and seed dormancy, so domestication depends of finding or inducing genetic variation for these traits. Hundreds of populations in several species have been collected without observing natural variation for seed retention. Differences in fruit morphology or seed retention have not been observed in recently collected populations of *C. tolucana* Peyr. (42 populations), *C. laminuligera* Koehne (8 populations), *C. leptopoda* Hems. (4 populations), and *C. viscosissima* Jacq. (50 populations). Continued exploration for seed retention variation is needed; however, finding seed retention variation in *Cuphea* is not a simple matter. Exhaustive germplasm collection and evaluation in promising species is needed for other equally important reasons and may result in the discovery of genetic variation for seed retention.

Another alternative, given the apparent lack of natural genetic variation for seed retention, is to attempt to induce genetic variation in traits controlling fruit development and seed dispersal. There are, as usual, no guarantees that useful mutants will be induced. Chemical mutagenesis of seeds to induce genetic variation. One or two generations of mutation breeding programs have been completed in various species. We have evaluated M2 populations of *C. lutea* Rose and *C. viscosissima* and generated M2 populations in *C. carthagenensis* (Jacq.) MacBride (n = 8) and *C. lanceolata* Ait. (n = 6) from ethyl methanesulfonate (EMS) treated seeds. Approximately 10% of the *C. viscosissima* M2 population of 1000 plants was comprised of macro-mutations. No seed retention mutants were observed. There were no macro-mutations in the *C. lutea* population. We presume macro-mutations are not observed in *C. lutea* because of duplicate genes and speculate that *C. lutea* is a polyploid, probably a tetraploid. Chromosome numbers in *C. lutea and C. viscosissima* are n = 14 and n = 6, respectively.

Seed dormancy in *C. viscosissima* forced us to evaluate a small M2 Population (1000 individuals) in 1988. M2 plants were derived by excising embryos from M2 seed, growing seedlings in the greenhouse, and transplanting to the field. Seed is dispersed in *Cuphea* through dorsal splitting of the floral tube and capsule and upward rotation of the placenta. Seeds are easily freed from the placenta. The fruit or capsule wall is extremely thin and does not protect the seed. The floral tube is sturdy but does not protect the seed once the placenta has rotated upward. The physical and physiological mechanisms leading to placental rotation, floral tube suture formation, and seed dehiscence have not been investigated.

Insect pollination is required in allogamous species, e.g., *C. laminuligera, C. lanceolata* , and *C. leptopoda.* A suitable pollinator for commercial plantings has not been found. Experimental plantings are mainly pollinated by bumblebees. The long floral tubes of allogamous species prevent honeybees from gaining access to the nectar. Bumblebees have long proboscises and are able to forage for nectar. Commercial production of the allogamous species is not possible without solving the insect pollinator problem. A simple solution seems unlikely. We are, therefore, discontinuing breeding activities in allogamous species. Breeding efforts in autogamous species have, until 1988, been hindered by limited germplasm resources.

C. lutea, C. viscosissima, C. tolucana, C. wrightii, and *C. carthagenesis,* among the autogamous species, have been most intensively investigated. *C. lutea* and *C. wrightii* (n = 22) are undoubtedly allotetraploids. Seed treatments with chemical mutagens in *C. wrightii,* like those in *C. lutea,* failed to generate macro-mutations. Thus, "observable" macro-mutations may not be generated, at least not at acceptable frequencies, in these species. The main criteria we are using to select among species, besides those criteria already mentioned, is adaptation and seed yield.

We have seed yield data on *C. lutea* and *C. wrightii* but not on *C. viscosissima* and *C. carthagenesis*.

Seed resources of the latter two species have been limited. Furthermore, seed dormancy has limited seed increases in *C. viscosissima* and prevented the establishment of *C. viscosissima* entries in yield experiments. Seed wields of *C. lutea* have exceeded those of *C. wrightii* in Oregon. Seed wields of *C. lutea* from swathing on tarps and subsequent threshing ranged from 400 to 1200 kg/ha at different harvest dates in 1987 at Medford, Oregon. There are two important points about these seed yields. First, experimental seed yields are undoubtedly substantially greater than those achieved by combining. Harvesting methods are under research and have not been standardized or optimized. These seed yield data are useful for comparing species but not for predicting yields under commercial cultivation. Second, single harvest seed yields represent about 10-40% of the seed yield potential of equivalent populations with seed retention.

Economical seed yields cannot be obtained without seed retention. The crop could be harvested after a killing frost and seed yield could accumulate to the end of the growing season if seed were retained. Thus, capturing seed yield is a problem but seed yield *per se* is apparently not a problem. *C. viscosissima* was grown in a yield trial for the first time in 1988. This species appears to have great seed yield potential, perhaps exceeding other species. A solution to the dormancy problem in *C. viscosissima* is undoubtedly forthcoming from selection programs and will greatly aid us in experimenting with this species. *C. carthagenesis* is not well adapted to cool climates; however, it is apparently well adapted to the warm climates of the southeast US. *C. calophylla* (n = 8), an uninvestigated species, may have an ideotype superior to *C. carthagenesis* and is worth investigating (Graham pers. commun.). Yield data on *C. carthagenensis* and *C. calophylla* from the southeast U.S. and other areas are badly needed to rank these species.

Cuphea germplasm resources have greatly increased over the last few years owing to collections made by the USDA-ARS and OSU. A recent USDA-ARS collection of 50 *C. viscosissima* populations made by W.R. Roath and M. Widrlechner (USDA-ARS, Ames, Iowa) is particularly noteworthy OSU and USDA-ARS germplasm resources include 2 *C. lutea*, 6 *C. carthagenensis*, 7 *C. wrightii*, 46 *C. tolucana*, 6 *C. laminuligera*, 8 *C. leptopoda*, and 42 *C. viscosissima* accessions tracing to distinct wild populations. There are about 30 *C. lanceolata* accessions. One traces to a wild population. The others are from European botanical garden collections and have uncertain histories.

C. lutea, *C. calophylla*, and *C. carthagenensis* germplasm resources are limited to two, two, and six populations, respectively, with no apparent within population genetic variation. These populations probably went through bottlenecks since they were collected prior to the development of a USDA germplasm maintenance program. Thus, there is a definite need for increased germplasm resources in these species. W.R. Roath, USDA-ARS, Ames, Iowa is planning another U.S. exploration for *C. viscosissima* and an exploration in Brazil in 1989. Numerous *C. carthagenensis* and *C. calophylla* populations should be found in Brazil.

The numerical abundance of accessions alone is not a useful measure of genetic variation in the *Cuphea* collection. There is, for example, tremendous within accession genetic variation in allogamous species and no or limited within accession genetic variation in autogamous species accessions collected before 1986.

Some of these differences may be attributed to how these populations were preserved and multiplied; nevertheless, there are gross differences in total within species generic variation. Mating systems are often correlated with species gene pool variation. Allogamous species generally have greater total within species genetic variation than autogamous species regardless of how the variation is distributed. There is some evidence for this in *Cuphea* and abundant evidence in other species.

Similar patterns of variation have been observed for economically important traits in pre-1986 *Cuphea* accessions. *Cuphea* presents the greatest domestication challenge among the new industrial oilseed crops. Whether or not it is successfully commercialized hinges on eliminating seed dispersal. The economic rewards for doing so are obviously great.

Apiaceae

The Apiaceae (formerly Umbelliferae) represent another option for domestic production of a lauric acid feedstock. Several Umbelliferae have seed oils, rich in petroselenic acid. Petroselenic acid, while not an item of commerce, may be fractionated into lauric and adipic acids (commercially important acids) through oxidative ozonolysis. Several Apiaceae have the advantage, compared to *Cuphea,* of having no significant production or harvest problems. One serious, disadvantage is the expense associated with fractionating petroselenic acid. Lauric and adipic acid prices determine whether or not it is economical to derive these acids from petroselenic acid. Abundant and inexpensive lauric and adipic acid feedstocks do not favor the commercial development of a petroselenic acid oilseed crop.

There are no agronomic traits preventing commercial production of dill, coriander, or fennel as oilseed crops. Commercial seed production for the seed trade is already practiced. Seed shattering is a minor problem in dill and coriander but has been observed in crops allowed to stand for prolonged periods after seed maturation. Seed loss to shattering is minimized by sound harvest management. Certain Apiaceae have immediate domestic production potential; however, their development as oilseed crops seems unlikely without high oil percentage cultivars and increased interest in petroselenic acid among industrial users of lauric and adipic acid.

Meadowfoam

Limnanthes species are typically collectively referred to as meadowfoam. The species of main interest in Oregon is *L. alba*. The primary gene pool of *L. alba* is comprised of *L. alba* var. *alba* and *L. alba* var. *versicolor*. *L. grascilis* var. *Parishii* is functionally part of the primary gene pool of *L. alba* because fertile progenies are produced from, and there are no meiotic irregularities in, interspecific F1 hybrids between these species. *L. alba* and *L. gracilis* var. *Parishii* are morphologically distinct species and have disjoint distributions. Practically, speaking, however, the genes of *L. gracilis* var. *Parishii* are highly accessible to *L. alba* and there apparently are no cytogenetic or reproductive aberrations in crosses between them; therefore, segregating populations behave and should be treated like populations derived from intraspecific crosses. It is not known if *L. alba* and *L. grascilis* var. *Parishii* hybridize under open field conditions, i.e., without manual pollination under controlled conditions.

Limnanthes is comprised of species with mating systems ranging from highly autogamous to highly, allogamous. The allogamous species are insect pollinated. Highly allogamous species, e.g., *L. alba* and *L. douglasii,* are characterized by abundant allozyme and economic and morphological trait variation, inbreeding depression and heterosis, high seed yields relative to autogamous species, and the requirement of insect pollination. Highly autogamous species (e.g., *L. floccosa* var. *floccosa*) have less allozyme and economic and morphological trait variation and lower seed yields than allogamous species, no inbreeding depression, and no insect pollinator requirement. On the one hand, allogamous species are preferred because their seed yields and genetic variation are substantially greater than those of autogamous species. On the other hand, autogamous species are preferred because they do not require insect pollination.

Honeybees have been used in commercial meadowfoam plantings. There has been speculation about reduced seeds yields in certain years caused by reduced bee activity during cool, wet weather. This speculation is based on field observations; however, year to year variation in seed yield and pollinator activity has not been documented. Honeybees are apparently unable to make honey from meadowfoam forage because of a lack of nectar, thus increasing production costs. Despite these problems there are no competitive autogamous species. *L. alba* and *L. douglasii* are logical candidates for domestication.

Natural genetic variation for seed retention was discovered in *L. alba* by W. Calhoun and J. Crane at Oregon State University in 1972. Seed retention variation has apparently not been found in *L. douglasii*. Thus, *L. alba* has a definite advantage over *L. douglasii* even though *L. douglasii* seed yields have been greater in certain environments. There is substantial genetic variation for quantitative traits in *L. alba* but it has not been adequately exploited. There are several reasons for this situation. First, there is no USDA program for meadowfoam germplasm collection, evaluation, and enhancement. Second, efficient breeding methods have not been continuously, applied to the improvement of *L. alba*. Third, the selection program at Oregon has been restricted to a narrow germplasm base, mainly an *L. alba* population derived from a few individual plant selections in PI283703 and PI283704 and a few interspecific hybrid crosses between *L. alba* and *L. floccosa* or *L. gracilis*. Fourth, the heterotic potential of *L. alba* has not been exploited.

The USDA *L. alba* germplasm collection at Pullman, Washington is comprised of 18 accessions tracing to wild populations. Data on within and between accession (population) variation are not available, at least there are no published data. Additional *L. alba* accessions are stored in the collection of S.K. Jain at U.C. Davis. There is a need to organize the germplasm resources of *L. alba* and other species, preserve them in a national collection, evaluate existing populations, and collect and evaluate additional populations. Significant increases in meadowfoam seed yields are needed over the next several years to commercialize meadowfoam.

An efficient route to achieving needed increases is a breeding program based on recurrent S1 family selection for increased seed oil yield. The rate of increase, besides being method dependent, is dependent on the magnitude of heritable variation in base populations, selection intensity, and selection criteria. Improved meadowfoam cultivars originating in the Oregon program apparently trace to public (USDA) accessions PI283703 and PI283704. Exposure of "selected" populations to populations besides those originating from these PIs may have occurred during their development. Thus, new variation may, have been inadvertently introgressed into these populations since isolation was rarely used in their development. The record is not very clear on this matter.

An increase in generic variation in base populations, beyond the variation currently being exploited, is needed to exploit the yield potential of meadowfoam. However, new, variation must be introgressed without seriously compromising population mean seed yields. An efficient way of doing this is to ensure that elite population gene frequencies are greater than exotic population gene frequencies in populations developed by intermating elite and 'exotic' populations. One method of doing this is to develop backcross populations using elite populations as recurrent parents. This is not practical in meadowfoam since there is no way to emasculate large numbers of plants. An alternative is to synthesize and intermate new populations using a greater proportion of elite germplasm. At least one generation of intermating is useful beyond the initial hybridization generation.

Even greater increases in meadowfoam seed yields may be achieved if breeding efforts are focused on exploiting the heterotic potential of *L. alba*. Heterosis has not been exploited to this point—quite the opposite. Jolliff et al. have striven to increase the level of self-pollination in *L. alba* populations by developing interspecific hybrids between *L. alba* and *L. floccosa* and selecting for autofertility and the *L. alba* phenotype. Their purpose has been to eliminate the need for insect pollination; however, increased self-pollination in *L. alba* has detrimental effects on quantitative traits, e.g., seed yield and fecundity.

It is impossible to determine whether or not self-pollination has been increased in populations selected for increased autofertility since outcrossing rates have not been estimated. It is questionable at best to pursue a strategy, for increasing self-pollination in a highly allogamous species subject to inbreeding depression when the main stumbling block to commercialization is seed yield. Seed yield and self-pollination rate are strongly negatively, correlated in *L. alba,* so they cannot be optimally simultaneously selected.

Heterosis is typically exploited through the use of hybrid (intraspecific) and synthetic cultivars. Genetic pollen control mechanisms for hybrid seed production, e.g., cytoplasmic-genic male-sterility, have not been discovered in *L. alba,* so there is no possibility for hybrid seed production in this species short of using gametocides to sterilize female lines. A suitable mechanism for exploiting heterosis in meadowfoam is synthetic cultivars. Their development, however, requires greater time and financial resources than needed for the development of open-pollinated cultivars. Meadowfoam is a potentially useful new oilseed crop but suffers because there are no established markets for its seed oils. Without markets this crop will remain a novelty.

Stokes Aster

Stokes aster is a source of epoxy fatty acids. Epoxy fatty acids are presently obtained by processing linseed and soybean oils because there are no direct commercial sources. Stokes aster, however, is a potential commercial source of epoxy fatty acids. It is remarkable that this crop has not been exploited. It has no serious competition among new temperate oilseed crops. *Vernonia galamensis* L., an epoxy fatty acid oilseed crop, is not adapted to the U.S. *Vernonia anthelminia* Willd. is adapted to the U.S. but has a serious seed retention problem.

Stokes aster is a perennial and apparently has few agronomic problems. Stand establishment is difficult and slow but no selection work has been done to increase seedling vigor and reduce seed dormancy. It is difficult to determine whether or not perenniality is an advantage or disadvantage. I perceive it as an advantage since perennials should have a significant role in sustainable farming systems. Stokes aster is indigenous to the U.S. and has been extensively collected. Approximately 20 accession reside in the USDA germplasm collection. There apparently is substantial genetic variation in these populations. There is, for example, extensive variation in seed retention.

Campbell has selected populations with closed bracts giving seed retention. The market potential for seed oils rich in epoxy fatty acids is between 45,000 and 90,000 t/yr. Stokes aster ought to be able to capture at least part of this market and is a crop waiting to be exploited. Sustained breeding work and industrial interest are needed to get Stokes aster commercialized.

Money Plant

Lunaria annua or money plant is a source of long-chain fatty acids similar to those of meadowfoam. It is a cruciferous biennial species and reportedly has tremendous yield potential. There unfortunately, are no published seed yield estimates for this species. In the Willamette Valley in Oregon it can be grown like winter rapeseed. There are annual mutants but their agronomic potential has not been estimated. Money plant and meadowfoam have identical market development problems. Basically, there are no established industrial markets for their seed oils. There apparently are no serious agronomic problems with *Lunaria*, e.g., seed shattering is not a problem, but the development of this crop is doubtful without a significant change in the present market outlook.

Sesame

Sesame (*Sesamum indicum* L.), thought to have originated in Africa, is considered to be the oldest oilseed crop known to man and is now grown in many parts of the world including the U.S. Sesame seed is an important source of edible oil and is also widely used as a spice. The seed contains 50-60% oil which has excellent stability due to the presence of natural antioxidants such as sesamolin, sesamin and sesamol.

The fatty acid composition of sesame oil varies considerably among the different cultivars worldwide. After oil extraction, the remaining meal contains 35-50% protein, and is rich in tryptophan and methionine. Seeds with hulls are rich in calcium (1.3%) and provide a valuable source of minerals. The addition of sesame to the high lysine meal of soybean produces a well balanced animal feed.

Total world production of sesame in 1986 was 2.4 million metric tons, 65% of which was produced in Asia. The U.S. is the largest importer of sesame, importing about 40,000 metric tons per year mostly from Mexico. Almost all sesame consumed in the U.S. is as a spice for food products such as hamburger buns and other bakery goods. Minor uses of sesame oil include pharmaceutical and skin care products and as a synergist for insecticides.

Although a major world oilseed crop, sesame is primarily grown by small farmers in developing countries in the southern latitudes. Crop development programs in these countries are either small or nonexistent, and little progress has been made during the past 20 years. A revitalization of sesame research using modern plant breeding knowledge and new technologies could be of great value in improving the crop.

Plant Breeding of Sesame

Sesame yields are highly variable depending upon the growing environment, cultural practices, and cultivar. Worldwide yields averaged about 340 kg/ha in 1986; however, yields as high as 2,250 kg/ha have been obtained in test plots in Texas. A major contributing factor to low yields in sesame is that the seed capsules shatter causing a loss of large amounts of seed, particularly when the crop is machine harvested. Sesaco Corporation in Yuma, Arizona, has developed semi-indehiscent commercial cultivar with yields ranging from 600-1600 kg/ha.

A considerable amount of mutation and plant breeding work has been undertaken in several laboratories throughout the world. Mutation breeding has been successful in producing generic lines with the determinate habit. Combining the high-yield trait with the semi-indehiscence trait could prove advantageous for developing a machine-harvested sesame crop. An early maturing line with 3 seed capsules per axil was selected from a composite population in California (Paul Brookhouzen, personal communication). Preliminary tests at a field nursery in Minnesota showed encouraging yields, thus demonstrating a potential for selection of lines suitable for a short growing season.

Sesame plants can be regenerated from shoot apical meristems and hypocotyl segments and grown to maturity in less than four months. Shoots regenerated via organogenesis from apical meristems and hypocotyl segments involves little or no callus issue production and variability in the progeny of regenerated plants is expected to be minimal.

Figure 1. Sesame plant regeneration in tissue culture. A. Shoot proliferation from hypocotyl segments. B. Plants regenerated from zygotic immature embryos.

Thirty-nine seeds from 12 regenerated plants were planted in the greenhouse. Seeds from these second generation plants were analyzed and showed no major variation in fatty acid composition (Table 6).

Tissue culture methods involving a callus phase or regeneration via somatic embryogenesis are known to produce stable variants. Somatic embryos directly from the surface of the zygotic embryos of sesame in culture.

Somatic embryo induction in six cultivars ('Aceitera', `Arawaca', `Turen', `Piritu', `Maporal' and `Inamar) varied from 50 to 100%. A large number of plants can be regenerated using such a system. To date, over 500 plants of five sesame genotypes have been regenerated. Their progeny will be screened for variation in characteristics such as seedling growth, vigor, placental thickness, capsule dehiscence, seed size, seed dormancy, yield, oil content and oil quality.

Table 6. Fatty acid composition of oil in R, seeds (second selfed generation from tissue-cultured regenerants) of sesame.

	Aceitera		Maporal	
Fatty acid	Control	R2	Control	R2
16:0	9.2	9.7-10.2	9.7	9.0-11.2
18:0	4.3	4.2-5.0	4.5	4.2-5.7
18:1	36.3	33.9	30.7	31.1-38.7
18:2	49.2	48.9-51.1	54.5	45.9-54.9

Callus cultures derived from cotyledons and hypocotyl segments were induced to produce embryos, although induction frequencies were low. Long-term callus culture systems would be useful for in vitro selection studies involving selective agents such as pathogenic fungal toxins, herbicides, and minerals. Several plants regenerated from these cultures showed morphological variations such as stem fasciation and differences in vigor and branching. Progeny seed will undergo further field evaluation and biochemical analysis.

Tissue culture methods can also be used to facilitate wide crosses using embryo culture techniques. Although conventional hybrid crosses between cultivated sesame and its wild relatives have been attempted, in most cases hybrids were difficult to produce. In preliminary studies, we cultured zygotic embryos at various developmental stages, and plants were regenerated from embryos obtained 15 days after pollination. Similar methods, could be used to regenerate plants from embryos generated from wide hybrid crosses.

Cruciferous Oilseeds

The diversity of form, usage and adaptability among *Brassica* species has generated wide interest in crucifers as potential new crops. Among these are various interrelated *Brassica species, Raphanus sativus, Eruca sativa, Crambe abyssinica* and others. Brassica oil known as rapeseed oil is traded world wide. Brassica seed low in glucosinolates and erucic acid, known as Canola, yields high quality edible oil and high protein seed cake. Some rapeseed cultivars are rich in erucic acid (22:1) a component of resins and lubricating oils for engines and steel manufacture.

The brassicas provide promising systems for combining technologies in cell and molecular biology with generics and breeding for the creation and improvement of new crucifer crops. Erucic and other fatty acid synthesis is under simple genetic control. The amenability of brassicas to protoplast fusion, plant regeneration and transformation has resulted in the development of stocks with specific antibiotic and herbicide insensitivity.

The development of rapid cycling populations of major crucifers has resulted in the formation of the Crucier Genetics Cooperative (CrGC), an organization sharing information and crucifer generic stocks globally Through the CrGC comprehensive mapping of RFLPs, isozymes, morphological and disease resistance markers are being facilitated.

Rapeseed

High erucic acid rapeseed (*Brassica napus* L.) has important industrial oil applications. However, the resultant meal contains high levels of anti-nutrient glucosinolates, which can also break down into bitter tasting, bad smelling compounds that can make feeds containing it unpalatable. In this experiment, we evaluated possible deleterious nutritional effects of the high glucosinolate meal in finishing cattle (700-1000 lbs) diets. Forty-eight steers were fed diets with the protein supplement comprised of 33, 67, or 1 00% rapeseed meal substituted for soybean meal. Data showed no differences in average daily gains and feed efficiencies. Carcass quality was better for steers fed the diet containing 100% rapeseed meal supplement, with those animals having the highest percent choice, whereas those fed 100% soybean meal supplement had the lowest percent choice.

In 1986, research was undertaken to determine the effect of irrigation timing and nitrogen on growth, yield and quality of spring rape (*Brassica napus* L. cv. Westar) under a furrow irrigation cropping system in Wyoming. Irrigation timing emphasized the stem elongation phase of crop development between bud formation and flowering. Nitrogen in the form of ammonium nitrate was applied at rates ranging form 39 to 302 kg N ha^{-1} depending on year. These rates were applied either spring (preplant incorporated) or split-applied spring and summer (preplant incorporated and top-dressed). Delaying irrigation to late stem elongation decreased yield compared to irrigation at the earlier bud stage. Increasing nitrogen rates increased yield. Split-application nitrogen treatments yielded more than spring nitrogen application treatments.

Crambe

Seventeen crop-related long term average weather variables were used to classify the climate for production of crambe (*Crambe abyssinica* Horchst.) for Indiana, Iowa, Kentucky, South Dakota to aid in selection of favorable sites for crambe establishment. The climate was classified into five climatic agricultural zones using Ward's clustering method and tolerance of cultivars to temperature and moisture was determined. Crambe growth was evaluated for potential production areas. 'Prophet' out yielded 'Meyer', 'Indy', C-50 and PI247310 in all climate types. Climate type 3, which includes northern Indiana, central and western Iowa, and southeast South Dakota, is most favorable for crambe production.

7

Cereals and Pseudocereals

Development of new grains and pseudograins requires a coordinated effort of individuals involved in marketing, processing and production. Triticale, the product of new crop research in the 1950s; quinoa, a new gourmet food in the mid-1980s; and blue corn, a potential new crop for the 1990s.

Triticale (x*Triticosecale* Widdmark) Triticale, the derivative of wheat and rye hybridization, provided the world with a glimpse of the future of agricultural technology in the 1950s. Triticale represented the first serious effort at synthesizing a new crop and received acclaim in the form of an endowed chair at the University of Manitoba in 1954. Trials initiated at Obregon, Mexico in 1963 were being jointly conducted at CIMMYT and the University of Manitoba by 1965.

At the outset of the program, triticale was intended to "compliment rather than replace other cereals". But inevitably, perhaps, triticale took on new meaning. Protein ash, and fat levels are frequently found to be higher than wheat. Because the world viewed triticale as a cereal grain it was immediately thrown in with wheat and barley in the marketplace. Charles Jenkins wrote, "There are those who look upon hexaploid triticale as another wheat. Since it does not behave like wheat it is often at a disadvantage when treated like wheat. As a grain, triticale had been thrust upon a marketplace in which it was in adapted to compete."

Alternative markets for triticale began to rapidly develop. In the 1970s triticale expanded as a fodder crop and as a supplement to bread flour. As a fodder, triticale has had good success both in North America and in Europe with over 400,000 ha (one million acres) planted in just the southern U.S. in 1988. Vigor and standability have been good to excellent when compared to wheat used as forage. As a flour, triticale has some disadvantages when compared to wheat. Lorenz et al. found the flour to be deficient in gluten quality and quantity with a high proteolytic activity. Such enzyme activity weakens the dough by protein hydrolysis.

Triticale flour was also found to be weak in gluten strength. This could be remedied by dough conditioning with paddles rather than hooks. Triticale also had a high sulphydryl content which further weakened the dough. Sulphydryl content varied from 1.2-16 ~moles/g for seven triticales while three wheats tested contained only 0.9 to 1.0 ~moles/g. The interest in triticale did not die with its seemingly poor entrance into world markets. Improved cultivars and a renewed market interest in whole grain and multigrain breads have continued to provide interest in triticale.

Lorenz and Welch found certain triticale cultivars useful for breads, white rye rolls, and noodles. Extrusion produced acceptable breakfast cereals and pancake mixes. Triticale provided a significant example of development without a comprehensive action plan to include production, processing, and marketing. Despite this, triticale is finding its market niche based upon its adaptability as a forage crop and its use in specially bakery products. Its adaptability to marginal environments has led to the expansion of its production in Eastern Europe with the greatest concentration being in Poland.

The response to triticale's slow growth is best summarized by John Hawthorne: "When a major development in world food supply appears in the offering, the time required for its development and evaluation can be quite long. While the experiments continue, there is a danger that over-optimistic rumors may exaggerate its potential and precipitate (by reaction) criticism based on misunderstanding of the real claims of the situation." The development of triticale will continue. Its place in the market may eventually rival the other common bulk grains but until that time, triticale has taken on the aspect of a specialized use crop.

Development of QUINOA (*Chenopodium quinoa* Willd)

In the development of new crops, especially for human consumption, we tend to find a tremendous competition for lower priced raw products. Several factors must be considered before such a new crop endeavor is instigated. A new crop program initiated at Colorado State University in 1982 involved an advisory panel consisting of growers, bankers, industrialists, extension and research personnel. The function of this panel was to determine the key objective and sequence of approach to a new crop. The consensus of this panel was that before development begins, a market evaluation should be made. Markets for edible oils, industrial oils, pharmaceuticals, natural rubber, paper pulp, gourmet foods and others were evaluated. Even in 1983, at the completion of this phase of the project, it had become evident that industrial crops for oil and other raw products appeared to have good growth potential. Other areas of growth included gourmet foods and bulk grains.

The development of new low-cost replacement grains did not appear feasible because of the lack of market channel development, the presence of farm support programs for existing crops, and a lack of sufficient research, in general, to compete with common grains. Limitations in industrial crops came not from marketing or production but from an inability to process the crops within an economically feasible distance from the sources of production. A similar scenario developed for edible oilseed crops. Lack of processing of the raw, product limited the development potential of these crops.

The development of specialized or gourmet foods was not as limited by the lack of industrial processing as industrial crops nor was it as hampered by preexisting market channels in common grains. In the United States, the best example is wild rice (*Zizania aquatica* L.) but there are others including: canola (being used as a salad topping), cherimoya, kiwifruit, and pistachio. A common feature of all of these crops is a small market in comparison to staples such as maize, wheat, or soybeans. A second common feature is the relatively higher gross return per unit of cultivated land.

Gourmet foods, being small, specialized markets, have a built-in production limit. This limit is set by the products in which the crops can be utilized. However, as processing of gourmet crops increases, new products can be formed which increase market size. Production could then be responsive to the need for new markets. Quinoa, as a food grain, has been recognized for centuries as an important food crop in the high Andes of South America. The very name quinoa in the Quechua and Aymara languages means "Mother Grain." Within South and Central America, two closely-related species, Canihua (*Chenopodium pallidacuale*) and huazontle (*C. nuttaliae*) are also utilized for food. The descendants of the Inca Empire, 8 to 10 million Quechua and Aymara Indians, still use quinoa as an important component of their diet.

Quinoa's food value in the developed world lies primarily with its seed. The seed is small (2-4 mm), circular and flattened on two surfaces. The pericarp contains a resin with 2 to 6% saponin which must be removed prior to consumption. Nutritionally, quinoa is high in protein. In trials at Colorado State University seed protein content varied from 12 to 18%), relatively high compared to conventional cereal grains. A more complete proximate analysis was obtained by Johnson and Aguilera. Of greater interest to individuals in the United States was the quality of that protein. Johnson and Aguilera completed a pattern of amino acid composition comparing the amino acid profiles of quinoa, wheat, soy and the FAO standards for human nutrition.

Quinoa's mineral content was scrutinized by Ballon who observed some additional health benefits. In comparison with wheat, barley and yellow corn, quinoa was found to be higher in calcium, phosphorus, magnesium, potassium, iron, copper, manganese and, zinc and was lower in sodium than the other grains. Preliminary data from Colorado trials has shown trends similar to those of Ballon but lacked the dramatic differences of quinoa and comparative grains.

Quinoa is versatile. As a whole pseudograin, it is cooked like rice after removal of the pericarp. Its flavor is generally regarded as nutty with a texture similar to North American wild rice. The grain has been used in soups, pasta, as puffed cereals, in extruded foods (in blends with corn and with oats), as desserts and side dishes. Its flour works well with wheat flour or grain or corn meal for breads and biscuits.

Colorado's successful entry in development of quinoa took an odd twist. Based upon the philosophy of the PPM system, marketing strategies were given the highest priorities. The difficulties of developing a market for a crop virtually unknown outside of the high Andes was felt to be too large a project either for a Colorado State University project under my direction or for the non-profit association, Sierra Blanca Associates. As a result, a joint venture was established. Sierra Blanca's President, David Cusack, joined with two other individuals Steven Gorad and Don McKinley to form Quinoa Corporation in 1983 to do market testing.

Quinoa Corporation imported quinoa from individual farmers in Peru and Bolivia to initiate test marketing in 1983. Steven Gorad went to South America and literally went from farmer to farmer purchasing quinoa for export to the United States. In addition to difficulties in obtaining salable quantities of seed, Quinoa Corporation also needed to initiate an educational program to educate consumers, to develop marketing strategies and to persuade retailers to surrender precious shelf space. Test markets were established in the Denver-Boulder area of Colorado.

Quinoa's relationship to the Inca Empire and its extraordinary food value in a limited health food market were emphasized. The health food market place was an ideal market because of the esteem given to nutritional quality and the willingness of patrons to try new foods and to pay higher prices. Sales of quinoa have shown continued growth from 1984 to the present. In 1983, 48 random samples of Quinoa ecotypes originating in

Equador, Peru, Boliva, and Chile were evaluated by Colorado State University. Few of the southern Bolivian and Chilean types survived in the short season environment of Colorado. Virtually none of the Peruvian or Equadorian types produced seed either due to daylength requirements or temperature stress.

The Chilean ecotypes have received a high priority since that time. Trials established in 1983 included three sites on the High Plains of Colorado, three sites in the San Luis Valley (Alamosa, La Jara and Center), one site at Fairplay, Colorado and one site at Olathe Colorado. Successful yields were only obtained at elevations between 2121 m (7000 feet) and 3030 m (10,000 feet) in elevation. Of the 48 entries, successful seed production and harvest were accomplished only in the San Luis Valley and Fairplay sites and only with four ecotypes.

Tetrazolium analysis of pollen for viability in 1984, concluded that elevations in Colorado below 2,000 meters resulted in complete pollen sterility under field conditions. With the major environmental limits defined, production practices became the next priority. Planting density studies, irrigation requirements, nitrogen requirements, and germplasm evaluation culminated in an extension publication by Johnson and Croissant.

Additional germplasm development continued with the release of a Chilean germplasm identified as 'CO407' in 1987. Preliminary on-farm yield trials were initiated by growers in 1984 with commercialization beginning in 1987. Commercialization was delayed because of the lack of a processing facility to remove the saponin-laden pericarp. Such a facility was provided by the Pillsbury Company and the growers in 1987. In 1988, a quinoa growers association was formed and a small processing facility developed for Colorado quinoa. Quinoa production costs have been reduced dramatically as producers have become familiar with the crop.

Currently, the North American Quinoa Producers Association is selling quinoa domestically as well as exporting to European markets. Colorado State University is assisting in the development of new, extruded quinoa products under the direction of Drs. Klaus Lorenz and Joe Maga. New product marketing is under the direction of Dr. Harry Kruekeberg.

Blue Corn (*Zea mays* L.)

Blue corn is another ancient crop which was developed primarily by the Indians of the North American Southwest. The amylase and soft starch found in blue corn and a rather unique flavor has opened markets for this specialty crop. These markets currently include blue corn tortillas, blue corn chips, pancake mixes, cornbread mixes and blue cornflakes cereals. Mexican restaurants and grocery stores are currently utilizing these products. Blue corn sales are showing continued growth. Additional product lines as well as increasing consumer acceptance have increased sales dramatically.

Grain Amaranth

The grain amaranths (*Amaranthus* spp.) are native to the New World. Pre-Columbian civilizations grew thousands of hectares of this pseudo-cereal. Some indigenous populations are said to have used grain amaranth, along with maize and beans, as an integral part of their cropping schemes. The Aztecs relied on amaranth seeds (or "grain") as an important staple.

The reasons for the decline of amaranth after the arrival of the Spanish are unclear. In an effort to explain the decrease in grain amaranth cultivation, fanciful myths have arisen. The mystery is especially intriguing when one considers that maize, with which amaranth co-evolved, was selected and developed into a major world crop. The small seed size of amaranth may have been a partial cause for the reduction in amaranth cultivation. A small seeded crop, such as amaranth, requires greater attention to detail in the early parts of the growing season than does a larger seeded crop, such as maize.

By the middle of this century, the cultivation of grain amaranth had declined to the point where it was grown only in small plots in Mexico, the Andean highlands, and in the Himalayan foothills of India and Nepal. Even now, there is evidence that some traditional farmers are abandoning the cultivation of local landraces of amaranth as they devote more of their land to high-yielding "modern" crops.

The cultivation of grain amaranth is now in the process of expanding in a number of countries. Over the past 15 years, there have been some well-executed projects in which researchers, farmers and food processors have invested imagination, time and money on this crop. Grain amaranth research at RRC focuses on three areas: germplasm preservation and utilization, the development of improved lines, and outreach and information dispersal. Over the past ten years, RRC has sent seed of segregating accessions and improved lines of grain amaranth to researchers in over 70 countries. Many of those countries now have internally-funded amaranth projects of their own.

China, and a USAID sponsored amaranth research project in Nigeria. Grain amaranth research has been served well by a six-year grants program that was administered by the Board on Science and Technology for International Development (BOSTID) of the National

Academy of Sciences. The BOSTID program has provided funding for research on nine projects in five countries (Mexico, Guatemala, Peru, Kenya and Thailand). Annual meetings of the grantees have allowed for an exchange of information on: the progress on germplasm evaluation, cultural methods, food processing, and nutrition. In addition, BOSTID has funded a quarterly scientific newsletter, published in Guatemala.

There has been increasing interest in grain amaranth in the international community. A number of conferences for the promotion of the crop have been held in 1977, 1979 and 1984. Conferences were hosted in Mexico in 1984 and 1986, with presentation of papers from individuals representing 19 different Mexican institutions. The range of subjects covered included ethnobotany, germplasm development, agronomic research, nutrition and processing. In 1987, an annual amaranth conference was initiated in the People's Republic of China to bring together researchers in over 22 provinces.

The American Amaranth Institute (AAI), which has been organized to help promote research and development, is working closely with many of the research institutions. The AAI is also working with Crop Improvement Associations to develop standards for amaranth seed certification. The synergism of the completed and on-going projects at RRC, the domestic and international institutions, and the AAI has provided a wealth of information and experiences for modern farmers to draw on as they initiate commercial production.

Farmers are developing innovative techniques to find ways to produce grain amaranth economically. As a result of their ingenuity, the supply of commercially available amaranth has increased to the point where several food companies are producing amaranth products which can be purchased in many stores in the USA. The rising demand for amaranth food products will require a substantial increase in amaranth production during the coming years. In 1988, approximately 1200 ha of grain amaranth were planted in the Great Plains of the USA, and 1000 ha were planted in the Sacred Valley of the Incas in Peru. In addition, we have received reports from collaborating researchers that there are substantial plantings of grain amaranth in Mexico, Kenya, and China to produce grain for export, as well as for internal use.

Germplasm

The genus *Amaranthus* includes approximately 60 species, most of which are widely dispersed weeds. The weedy amaranths are referred to as "opportunists" because of the fact that they thrive in disturbed soils and tend to be associated with agricultural practices. The tradition of using some species of feral amaranths as a forage for livestock, as greens for human consumption, and as dye plants, continues in many places.

There is no clear dividing line between a "grain" type and a "vegetable" type. The leaves of amaranth plants which are being grown specifically for grain can serve as a food source, if the leaves are consumed when the plants are young. A few species which are grown primarily in the hot, humid tropics are prized for their high quality leaves. There are many cultivars of the species *A. tricolor* L. which are widely dispersed and cultivated throughout Asia and the South Pacific. The leaves of *A. dubius* Mart. ex Thellung are considered to be a delicacy in many areas of the Caribbean. A dark-seeded strain of *A. cruentus* L. is commonly cultivated as a vegetable in West Africa.

Amaranth which produce a large amount of biomass in a short period of time, can be used as a forage crop for domesticated animals. In China, amaranth has been cultivated expressly for use as a forage for cattle. There are several cuttings made per growing season. In Peru, the stover is grazed or milled for use as a feed supplement after the seedheads have been hand-harvested. Some types of amaranth accumulate toxic levels of oxalate and nitrate when grown under conditions of stress. Additional research is needed to better define the feeding value, potential toxicity problems and the most appropriate type of amaranth for use as a forage.

There are three species of the genus *Amaranthus* which produce large seedheads of edible, light-colored seeds. *A. cruentus* L. and *A. hypochondriacus* L. are native to Mexico and Guatemala. *A. caudatus* L. is native to the Andean regions of Ecuador, Peru, and Bolivia. Cultivated land races of the three species have a wide degree of genetic diversity which can be utilized to develop selections of improved lines with suitable characteristics for modem grain production.

Amaranth germplasm is catalogued and stored in germplasm banks in 11 countries. A descriptive taxonomic key for all the cultivated species of the genus *Amaranthus* was developed by Feine-Dudley. Since 1977, RRC has maintained a working germplasm collection. The 1400 accessions in the amaranth germplasm collection include representatives from 12 amaranth species. The collection includes germplasm from banks in other countries, material from germplasm collection trips which were made by RRC staff members, and materials which have been donated by collaborating researchers. The germplasm is catalogued according to "grain type".

The grain type categories may actually be a regrouping of the germplasm according to landrace. The grain type categories have proven to be useful to manage the huge amount of variability that exists within each species. Descriptions of all species and the grain types can be found in the RRC germplasm catalogue. A majority of the accessions in the RRC germplasm collection were collected as mass selections or single plant selections of cultivated landraces of those species which are commonly grown for their light colored seed. Many selections have been made from the segregating accessions in an effort to create uniform lines.

Seed from the amaranth accessions has been distributed to thousands of researchers and farmers around the world. The germplasm collection is the backbone of our varietal improvement program which is aimed at developing agronomically acceptable lines using classical plant breeding and selection methods. Germplasm characterization is being conducted at a number of locations around the world. Extensive, well-documented amaranth germplasm characterization has been conducted by organizations in India, Peru and Mexico. Germplasm from their collection will be entered into the Plant Introduction System for distribution.

Since 1982, the staff at RRC has collaborated with researchers who are conducting observations on 14 selected grain amaranth accessions (representing grain types from the species *A. hypochondriacus L., A. cruentus* L., *A. caudatus* L., and *A. hybridus* L.) to collect information on their performance at distinctly different climates and latitudes. Information generated at each location serves as a starting point to help determine which sources of germplasm should be exploited to develop improved lines of grain amaranth for any given area. Similar observations have also been made to characterize germplasm which was collected throughout India. Populations of landraces which were collected from

Mexico to Argentina have been observed for their genetic structure.

Germplasm enhancement has been conducted at RRC since 1977 using recurrent single plant selection and mass selection. Selections continue to be made for the development of ideotypes which meet the needs of modern agriculture.

The first released line of amaranth named 'MT-3' was announced by the Montana Agricultural Experiment Station (Cramer 1988). It was selected from a segregating accession of *A. cruentus* L. originally collected in Mexico. Similar programs are presently in progress in Mexico and Kenya. The University of Cuzco in Peru has released three cultivars of *A. caudatus* L. which have been named 'Oscar Blanco, 'Noel Vietmeyer' and 'Alan Garcia'. These three cultivars are being grown commercially on hundreds of hectares by farmers in Peru. A cultivar of *A. hypochondriacus* L., named `Annapurna' has been released in India.

The first released cultivars have helped to promote the commercialization of this crop. However, the gene pools of all of the cultivated amaranth landraces lack one or more desirable traits which are perceived as requirements for commercial production. For example, with adequate moisture and fertility, all of the improved varieties which mature and produce good seed yields of light-colored grain at northern latitudes reach a height of 2.5 m at maturity, too tall to harvest with modern grain combines.

Genetics and Plant Breeding of Amaranthus

The genus *Amaranthus* provides fascinating material for geneticists to tackle. Amaranth is a predominantly self-pollinated crop, with varying amounts of outcrossing. By growing amaranth in isolation, it is possible to control the amount of outcrossing for the development of true-to-type lines from segregating accessions within only a few generations of selection. Basic genetic studies were initiated in the late 1950s. Chromosome counts were made. Polyploids were also identified. Studies have been conducted to document the amount of variation within Indian populations of amaranth. The possibilities for interspecific hybridization have been documented.

The cultivated species of grain amaranth are monoecious. The basic techniques for making emasculations and pollinations have been described and refined. The potential for the development of hybrids has been advanced by the identification of gene markers in the program at Davis, and other locations. Plant and seed pigment patterns, leaf characteristics, time of flowering, and determinate vs. indeterminate growth, have been identified. An additional marker that has been observed in breeding trials at RRC is bract length. The perisperm and, starch of the grain can be used as potential gene markers.

Markers can be used to identify and discard self-pollinated plants in the F_1 generation. In addition, markers allow breeders of grain amaranth to avoid the process of emasculation when making crosses. The identification of three sources of cytoplasmic male sterility paves the way for additional hybridization techniques. Genetic investigations have been conducted in India to learn more about the traits which control yield, including harvest index, yield per plant and weight per 1000 seeds.

There have also been some unique investigations to identify the mode of inheritance of traits that affect the nutritional characteristics of the grain. Work in Japan has focused on the identification of factors related to the unique starch characteristics of the grain. In India, the mode of inheritance of grain protein percentages has been studied. In 1978, an applied amaranth breeding program was initiated at RRC to develop improved plant types for mechanized production in the USA and Canada. Using classical plant breeding techniques, we have developed a series of advanced breeding lines (F6-generation, and higher) which are being grown at experiment stations and on commercial farms in the USA. The lines are being distributed as part of the "K-series."

One of the lines, 'K-343', is the progeny of a cross of a white seeded *A. hypochondriacus* L. from Mexico and a black seeded *A. hybridus* L. from Pakistan. This single line was grown on most of the commercial plantings of grain amaranth in the Great Plains of the USA in 1988. It has an unbranched short plant habit at high plant densities, and slightly earlier maturity, as compared to 'MT-3' and other selected Mexican landraces which have been grown by farmers. The current concentration on one source of germplasm has the potential to provide a serious setback to the development of the crop. Additional resources are needed to initiate and expand applied breeding programs to meet the needs of the anticipated expansion in grain amaranth production.

The initiation of the seed certification process which was begun in 1988, will help to address the need for sources of quality seed of improved lines of grain amaranth. The future of amaranth is dependent on the careful recombinant selection of genotypes from the germplasm collections. We have had to use unusual sources of germplasm to find traits such as early maturity and shorter plant stature. The ideotype required for mechanized production in the USA has a number of characteristics in common with the landraces from Mexico. We have used highly branched sources of dark-seeded *A. cruentus* accessions ("African" grain type) from West Africa as a source for early maturity. A source for short plant stature was found among a few highly branched, shattering, dark-seeded *A. hybridus* L. ("Prima" grain type) accessions from Asia.

In addition, some traits are affected by environmental influence, such as plant height, days to maturity, plant architecture, and drydown. For example, accessions of *A. hypochondriacus* L., the Nepal grain type, when grown in Kenya (at the equator and in a semi-arid environment), produces short plants (<1 m), with multiple flowering stems, and matures in 60 days with a high harvest index. When the same accession is grown in Pennsylvania, the plants develop a tall plant (>2 m) with a single stem, matures in >160 days and has a low harvest index.

Additional research needs to be conducted to develop information on the barriers that prevent interspecific hybridization. In general, there seem to be few barriers that exist between crosses of *A. cruentus L., A. hypochondriacus,* L. and *A. hybridus* L. However, crosses with *A. caudatus* L. and any of the above species often result in non-viable progeny. Some grain types within the species *A. hypochondriacus* L., when crossed by *A. caudatus* L. behave differently from others. For example, we have obtained viable F1 progeny from a cross of *A. caudatus* L. by accessions from the grain types "Aztec" and "Nepal" of the species, *A. hypochondriacus* L. However, crosses of *A. caudatus* L. by accessions of the grain type "mercado" of the species *A. hypochondriacus* L., have produced very few viable F1 progeny.

To assist in the development of selected progeny from breeding programs, RRC has distributed segregating material from F_2 and F_3 generations of our breeding progeny. The segregating material is grown at the location where the advanced lines will be grown to allow for selection pressure at that location. This technique is currently in progress in Peru, Thailand, China and Kenya, where new progeny are being selected. To date, the most inclusive evaluations of segregating progeny of interspecific crosses have been conducted in Peru.

Most of the first progeny of interspecific crosses observed in Peru did not perform well. However, a few accessions of *A. caudatus* L. crossed with two accessions of the Aztec grain type of *A. hypochondriacus* L. produced progeny that included segregates which were shorter and earlier maturing than the commonly available varieties of *A. caudatus* L. that are presently being grown in Peru. The performance of the segregating progeny and the selections that have been made will provide important clues to help select the most appropriate sources of genetic material to use in future breeding programs in Peru. Similar approaches will be useful at other locations.

Amaranth stands to benefit from some of the new techniques which are being developed in molecular generics. Callus formation has been documented which will allow in-vitro cultivation for advancement of breeding lines. The gene expression of certain carboxylases is also undergoing continuing studies. A project has been initiated in Mexico by Herrers in which gene expression is investigated.

Use of Grain Amaranth

The food value of grain amaranth was recognized by people from Mexico to Peru to Nepal long before any, nutritional analyses had been conducted. Because it is easy to digest, amaranth is traditionally given to those who are recovering from an illness or a fasting period. In Mexico, grain amaranth is popped and mixed with a sugar solution to make a confection called "alegria" (happiness). A traditional Mexican drink called "atole" is made from milled and roasted amaranth seed.

In India, *A. hypochondriacus* L. is known as "rajgeera" (the King's grain) and is often popped to be used in confections called "laddoos," which are very similar to Mexican "alegria" In Nepal, amaranth seeds are eaten as a gruel called "sattoo" or milled into a flour to make chappatis.

During the last ten years, a number of overviews have been published which provide a wide range of information on the nutritional components, digestibility and potential problems that will be encountered by those who intend to use grain amaranth as a food product. There are several hundred other literature citations that will be included in a bibliographic summary to be entitled "Human Nutritional and Food Applications of Grain Amaranth".

The most studied nutritional aspect concerning the food value of grain amaranth is the identification of the limiting amino acids of the protein component. The crude protein content of selected light-seeded grain amaranths has been reported to range from 12.5 to 17.6. Amaranth grain is reported to have high levels of lysine, a nutritionally critical amino acid, ranging from 0.73 to 0.84% of the total protein content. The limiting amino acid is usually reported to be leucine, although some reports indicate that threonine actually may be the amino acid which is more biologically limiting than leucine.

The potential complimentary nature of amaranth protein has been studied by combining amaranth with wheat, sorghum and maize. Ordinary maize meal supplemented with as little as 12.7% (by weight) of toasted amaranth flour provides a nutritionally superior source of protein that can satisfy a good portion of the protein requirement of young children, and provide approximately 70% of diet energy. A combination of rice and amaranth in a 1:1 ratio has been reported to approach the FAO/WHO protein specifications.

The starch component of amaranth is distinctive. The starch granules are polygonal, measure 1 to 3 mm in diameter, and have a high swelling power. There is a distinctive gel characteristic to the starch. Waxy and non-waxy starch granules have been identified. Interest has been expressed in specialized food and industrial applications for amaranth starch as a result of its distinctive characteristics. Amaranth grain contains 6 to 10% oil, which is found mostly within the germ. It is predominantly an unsaturated oil (76%) and is high in linoleic acid, which is necessary for human nutrition.

Due to the fact that grain amaranth has high protein, as well as a high fat content, there is the potential to use it as an energy food. Using milled and toasted amaranth products, digestion and absorption was found to be high in human feeding studies. The balance of carbohydrates, fats, and protein, allow amaranth the opportunity to achieve a balanced nutrient uptake with lower amounts of consumption than with other cereals. It has been noted that high protein rice is the only other cereal which has been cited to satisfy protein and energy needs.

Animal feeding studies indicate relatively high protein qualities. However, in some studies, weight gains were much lower than would have been expected for reasons that are not clear. The digestibility and the protein efficiency ratio are improved if the grain is heat processed. The removal of lectins by heat processing has been reported to improve the protein efficiency ratio of the amaranth flour. There are a number of viable methods for processing, including popping, flours milled from toasted grain, heat-rolled flakes, extrusion, and wet cooking as a gruel.

An interesting application for amaranth is to use it as a food for people with allergies to other grains. The seed of "grain" amaranth is not a grain from a cereal plant, but is rather a pseudocereal from a dicotyledonous plant. It is unrelated to any other food crops that are commonly consumed, which makes it less likely to cause problems to people who have built up allergies due to repeated consumption of the same foods. Grain-free recipes which include amaranth flour have been published.

In the USA, many amaranth products are being produced by specially companies which cater to the health-conscious market. Amaranth has been successfully processed in combination with other grains to produce cold, breakfast cereals. It is also being used for mixes that are used to prepare hot breakfast cereals and pancakes. In addition, there are breads, crackers and pastes on the market. Popped amaranth grain continues to attract considerable attention. The popped grain provides opportunities for processors to develop innovative products. Grain amaranth is a new crop that is in its adolescence. The cultivation and utilization of grain amaranth will continue to increase as more information is developed to exploit the market niches for high quality protein foods.

Perennial Grain Crop

Perennial grains can provide advantages over annuals including lower production inputs as well as year round soil protection. Mechanical field operations in a perennial grain system will be greatly reduced because the soil will not be re-worked each year, thus saving on fuel and labor costs. Seed purchases and planting would be done every five to ten years rather than every year as with annual grains. Where legumes are included with plantings of perennial grains, the need for commercial fertilizers could be reduced.

A perennial grain/legume mixture could be considered analogous to a grass/legume pasture. However, instead of the primary harvest product being the hay as in a pasture, the primary harvest product will be the grain from the perennial grass. In some areas, the perennial grain/legume fields can be used for hay production or grazed as pastures after the yearly seed harvest. In addition, the use of perennial grains could rebuild soil structure through increased soil organic matter, water infiltration and biological activity in the soil. Because the soil will have vegetative cover during the entire year, wildlife such as game birds and mammals could also benefit.

Intermediate wheatgrass is a cool season, rhizomatous perennial which is used extensively for pasture, hay and soil erosion control in the Great Plains and intermountain West. Since 1986, a total of 250 accessions of intermediate wheatgrass have been acquired by the RRC from a variety of sources. Half of these were obtained from the USDA Plant Introduction Office in Pullman, Washington. These PI accessions were originally collected in the Soviet Union, Iran, Turkey and other areas of the eastern Mediterranean region. An additional 100 accessions were collected as single plant selections from breeding lines growing in field plots at the Northern Great Plains Research Center in Mandan, North Dakota.

This germplasm is being tested in small plots at the RRC to determine seed production capabilities. Seed yield components are being evaluated for each accession in order to select high yielding lines for future breeding programs. Characteristics being studied include seed size, measured as weight of 100 seeds, number of fertile florets/seedhead, a measure of seedhead fertility and fertile tillers/plant. Both seed size and fertility are highly heritable characteristics in intermediate wheatgrass. Therefore, clonal selection based on these two yield components should be effective in increasing seed yield.

The nutritional qualities of this perennial grain are similar to wheat and rye. Compared to wheat, intermediate wheatgrass has a slightly higher protein level with a higher content of the sulfur containing amino acids, methionine and cysteine. This could be valuable for complementation with legumes in food applications. Flavor qualities of intermediate wheatgrass grain are also good. At the 100%, 66% and 33% wheat flour replacement levels, whole flour from intermediate wheatgrass grain performed well in baked products such as muffins and rolls. Wheat flour-perennial grain flour blends were preferred (sweet/nutty) over the wheat flour control.

Farinograph tests of whole grain flours made from several different cultivars of intermediate wheatgrass were conducted by Tracy Bergman at the WRRC. The performance of 'Luna' was very similar to that of a high quality dough forming wheat. Interestingly, 'Oahe' did not compare as favorably. This demonstrates that th e re may be sufficient variability among the intermediate wheatgrasses to select and develop lines for different culinary purposes as in wheat.

Herbicide-tolerant Germplasm in Rice

Red rice (*Oryza sativa* L.) and cultivars of white rice belong to the same genus and species. Red rice is one of the most common and troublesome weeds of rice in the southern U.S., causing an estimated loss of $50 million dollars annually. Red rice is a problem weed because it has a red pericarp or seed coat which is commercially unacceptable; reduces yield because it competes for light, nutrients, and space; tillers more profusely; increases lodging because of its weak straw; reduces milling yield because it requires more severe milling to remove the red pericarp; and shatters easily which makes it hard to eliminate from commercial fields.

Control of red rice in commercial rice fields is economically prohibitive and selective herbicides for control of red rice in rice are not available. Strawhull and blackhull are the two principal types of red rice. Chemical control of red rice in commercial fields has not been achieved because of the similar biological and physiological properties of red and commercial rice. However, rice rotated with upland crops combined with effective herbicides will control red rice. Soybeans grown for two years and followed by rice the third year has reduced red rice infestations sufficiently for satisfactory rice production. Alachlor, which is absorbed by germinating seeds, controls red rice in soybeans.

The development of rice cultivars which are tolerant to a herbicide that controls red rice would help control red rice in commercial rice. Intraspecific variation in tolerance to herbicides has occurred in crops such as barley (*Hordeum vulgare* L.); oats (*Avena sativa* L.); corn (*Zea mays* L.); sorghum [*Sorghum bicolor* (L.) Monech]; sugarcane (*Saccharum officinarum* L.); soybeans [*Glycine max* (L.) Merr.]; potato (*Solanum tuberosum* L.); and tomato (*Lycopersicon esculentum* Mill.). Rice cultivars and advanced lines that are tolerant to molinate, an herbicide which controls barnyardgrass, have been identified.

8

Germplasm and Breeding

Germplasm is a term used to describe the genetic resources, or more precisely the DNA of an organism and collections of that material. Genetic resources encompass all the various forms of germplasm that are available for collection, storage and use. The genetic diversity of crops, represented by traditional local cultivars and wild relatives has been disappearing rapidly during recent decades. Concerted international activities to collect and preserve crop genetic resources were initiated by the Food and Agriculture Organization (FAO) of the United Nations. Technical conferences organized by FAO in 1961, 1967 and 1973 stimulated the global awareness with regard to conservation of genetic resources.

In 1971, the Consultative Group on International Agricultural Research (CGIAR) was formed with cosponsorship from FAO, the United Nations Development Programme (UNDP) and the World Bank. The 1973 FAO Technical Conference and a United Nations Environment Conference in Stockholm in 1972, led to recommendations for a global programme. The International Board for Plant Genetic Resources (IBPGR) was subsequently established in 1974 by the CGIAR as one of 13 International Agricultural Research Centres (IARC).

As a secretariat organization, IBPGR was charged with salvaging threatened germplasm and ensuring that it was described, documented and conserved. Before IBPGR was created, genetic resources programmes were based in the developed world or in other centres of the CGIAR. Many of the existing programmes were the result of plant introduction programmes or were centered around historic collections. In 1974, fewer than five conservation facilities worldwide were available for good storage of seed. Also, collections of vegetatively propagated crops were small, did not represent existing variation in local varieties and had suffered severe generic losses due to handling . The documentation of both seed and vegetative collections was inadequate or non-existent. There was a significant lack of information on the ranges and patterns of variation in field populations of old cultivars, landraces, weedy and wild relatives, and the degree of threat to these populations.

IBPGR plays a catalytic role in establishment of research programmes and projects by providing initial funding and scientific expertise. It has a responsibility to assist in establishing new genetic resources programmes and to integrate these with current ones as quickly as possible. Very early, a strategy was developed that included the collection, conservation, characterization, and documentation of generic resources. This ran concurrently with the initiation or strengthening of national genetic resources programmes and training in and for developing countries. The IBPGR has worked with and collaborated with almost all countries of the world regardless of political affiliation.

The activities of the IBPGR touch all areas of plant genetic resources work. The field programme is served by both headquarters and field staff and includes acquisition, conservation and documentation of genetic resources. IBPGR has sponsored over 400 collecting projects in all parts of the world (115 countries) that have resulted in over 166,000 samples deposited in genebanks. Because IBPGR does not maintain any, germplasm, it enters into agreements with national and international genebanks to conserve collected germplasm This has resulted in a network of over 30 base (seed) collections.

Assistance has been provided whereby, 15 long-term storage facilities and numerous short-term facilities have been established, many in developing countries. Support has also been directed toward establishment of field genebanks in some areas. Standards for conservation of seed have been developed and accepted throughout the international community. Characterization and preliminary evaluation of genetic resources is also an active area. Descriptor lists have been or are currently being developed for over 60 crops or crop groups. Support has also been provided for characterization and multiplication of some collections.

Documentation and information management activities have included providing and developing directories of institutes holding various material, technical assistance to centres in developing countries with major collections and development of standards for the establishment of crop databases. Funding for the establishment and compilation of crop databases has been provided for a number of the major crops. Training has been partly or completely funded by IBPGR for personnel in technical, management and supervisory areas. A variety of short technical courses have provided training to over 600 trainees from developing countries with over 100 receiving extended postgraduate training financed by IBPGR.

The research programme of IBPGR is a relatively new area within IBPGR. There are three areas of potential emphasis. Research in conservation technology is still needed, for seeds and vegetatively propagated crops through in vitro techniques. A major project involves investigating the logistics of conservation of in vitro material and the genetic changes that occur in cassava during in vitro storage. Pathological aspects of germplasm conservation and exchange is another major area. A conceptual framework has been developed for an enclosed quarantine system, a component of which is disease indexing.

Genetic diversity research is needed because of the constraints in implementing the collecting of diversity from wider gene pools. Current research is based on species complexes in primary, secondary, and tertiary gene pools depending on the sophistication of breeding specific crops. Throughout the implementation of the above activities, IBPGR has developed a large consortium of collaborators throughout the world. It has developed into a repository of information based on current and past contact with hundreds of institutes and individuals worldwide.

Domestication of New Crops

Domestication of wild species for development as new crops have been attempted for several species in many different programmes. Significant results have not been obtained in most cases. Many of these crops show potential but the ultimate success is determined by market forces and only sugarbeet, rubber and oil palm have been domesticated in recent times. Science provides a great potential for the rapid exploitation of wild plants, however economics demands rapid exploitation of those species that can provide compounds needed for pharmaceutical, cosmetic or insecticidal applications.

Though necessary, these plants are not of high priority with IBPGR.

In situ conservation efforts with special regard to natural reserves have received much attention. These areas permit unhindered evolution and may provide the best alternative to conservation of these wild plants. The IBPGR has established links with the World Wide Fund for Nature (WWF), the International Union for the Conservation of Nature and Natural Resources (IUCN), and the Man and Biosphere Programme of Unesco with the aim of stimulating in situ conservation. These areas are concentrated mainly in the tropics where the widest range of potential new crops occur.

IBPGR activities have covered a wide range of crops and crop groups. Subtropical and tropical fruits are an important asset to the countries in which they are produced, both nutritionally and economically. The IBPGR has provided support for the collection, conservation, ecogeographic and systematic surveys of fruits that are regionally and locally important. Current efforts emphasize mango, citrus, banana and related wild species, especially those in Southeast Asia. The biological aspects of the conservation and propagation for banana and citrus lead to the support of research to develop in vitro tissue culture techniques for conservation and collection.

IBPGR has been involved in the collection and conservation of many different tropical and subtropical fruits that are underutilized or unexploited in the U.S.:

Passiflora edulis (passionfruit) has become increasingly popular as an ingredient in fruit drinks, but at least 40 other species of *Passiflara* have edible fruits, between ten to thirteen of these are reportedly cultivated. Most of the cultivated species are naive to South America but both the cultivated and wild species are reported to naturalize easily even tending to weediness. While many of these species grow throughout the tropics, most are consumed locally with only a few traded internationally. There is the potential to utilize fruits in products such as jams, jellies, ice creams and other desserts and one is important medicinally (*P. incarnata*).

Annona cherimola (cherimoya) is a backyard crop native to the nonhem Andes, the fruit of which has slightly acid but delicate banana-pineapple flavoured flesh. The crop is in increasing demand in the world market and is exported by a few countries including Spain and Chile. It has been used as a fresh fruit and in ice creams or deserts. There are some small collections of *Annona* germplasm. A relatively large colleclion of *A. muricata* is located in Mayaguez, Puerto Rico.

Solanum muricatum (pepino), produces a seedless, watery and pleasantly favourable fruit on a very fast growing shrub. In its native South America, it is a traditional backyard crop. The fruit is also produced on a small scale in Hawaii, southern Califomia and the southeastem U.S. It is currently commercially produced in New Zealand and has potential for extended commercial cultivation as a specialty fruit in the U.S.

Cereals and Grain Legumes. Since its creation, the IBPGR has given emphasis to the major cereals because they are staples for much of the world's population. IBPGR has organized the collection of landrace and wild material, sponsored duplication and conservation, and provided funding for characterization work on maize, wheat, barley, rice, sorghum and millets. Ecogeographic surveys of the Triticeae tribe, wild Sorghum, Eleusine and teosinte have been supported in areas of high diversity.

Several species of the *Pennisetum* millets are very valuable human food in and portions of Africa, Asia and the Mediterranean. These species, probably of central African origin, thrive in ecologically marginal lands, including those with poor soils and areas where drought is common. The IBPGR has been active in collecting substantial landrace material, most of which has been the common pearl millet (*Pennisetum glaucum*). The relatively high average protein content of the grain and the potential to use the plant as silage in somewhat unproductive lands creates interesting opportunities for developing *Pennisetum* species as new crops in the U.S.

Food or grain legumes are a major staple in Latin America, Southwest Asia, the Indian subcontinent, East Asia, and Africa. Legumes are among the best natural sources of vegetable proteins, calories, vitamins and minerals. Some are also high in oil. Many different legumes have received considerable attention with regard to germplasm acquisition and conservation within the IBPGR programme. *Phaseolus, Vigna, Arachis, Cicer, Lens, Pisum, Cajanus, Vicia* and *Glycine* have been collected and conserved in many regions of the world. IBPGR works closely with many national programmes and CIAT, ICARDA, ICRISAT, the International Institute for Tropical Agriculture (IITA), and the Asian Vegetable Research and Development Center (AVRDC) in a concerted effort to collect, conserve and characterize legume germplasm.

Two genera of food legumes, *Lupinus* and *Psophocarpus* are not currently exploited in the U.S., but may have potential sources as interesting crops. The lupins have two centers of diversity, one in the Mediterranean and one in South America. Many lupins are very vigorous, rapidly growing on poor land, with easilv harvested seeds high protein and oil. They have been used as grain food, animal forage, for soil improvement and as ornamentals.

The Latin American species, *Lupinus mutabilis,* is the most important cultivated lupin in South America because of its wide use as food in the Andean region. It has been collected extensively and conserved throughout its region of diversity for many years with IBPGR support. The Mediterranean lupins (including *L. albus* and *L. luteus*) are not as important there as *L. mutabilis* is in the Andes. The IBPGR has been actively supporting national institutes, especially in Portugal and Spain, to collect these and related species, principally in the Iberian peninsula. The high alkaloid content in the lupin seed creates a barrier to their quick acceptance. It is possible, however, to change this through breeding, potentially releasing a vigorous plant with a high protein content.

The winged bean has been considered a potential new crop in the U.S. for many years. *Psophocarpus tetragonolobus*, a native of tropical Asia, is a very important crop in Papua, New Guinea and South east Asia because of its edible leaves, seeds, pods and tuberous roots, all with a high protein content. The beans can also produce an edible oil. There are currently large holdings of this species, formed principally through IBPGR support of many different collecting missions. These germplasm collections can facilitate research on the adaptability of the species to northern latitude daylight and climatic conditions. The potential use of the winged bean as a high protein forage in conjunction with harvestable tuberous roots, or as an edible legume seed creates an alternative to some of the more traditional U.S. bean and root crops.

Quinoa produces a grain on a broad leafed plant. It is rugged, heavily seeded and thrives in poor soils. The grain is rich in protein, contains a good amino acid balance, and possibly can be improved to be a better protein source than most of the common cereals. Quinoa has a light taste, may be used to thicken soups, or popped like popcorn. A detriment to its use is that the outer lavers of the seedcoat of most types contain saponins (water soluble glucosides) that cause it to be bitter-tasting. This can be removed through breeding, washing, or milling. The grain can currently be purchased inside the U.S. and is grown successfully in Colorado and England. With several related cultivate species (*C. pallidicaule, C. nuttaliae*) and wild species there is a broad genetic base that can be used for improvement and adaptation of quinoa to non-native regions.

The amaranths have received significant world-wide attention because they are fast growing and produce high-protein grains in large seed heads. Vegetable amaranths are widely known for their protein-rich leaves that can be harvested many times a year. The IBPGR has been active in the collection, conservation and characterization of the genetic resources of many species of amaranths. *Amaranthus hypochondriacus, A. cruentu*, and *A. caudatus* are native to Mexico, Central America and the Andes, and are important in these regions but also in Africa and India principally as a grain crop.

The high lysine grain is usually parched and milled for dough for pancakes, cooked for gruel or made into drinks. The vegetable amaranths (*A. tricolor and A. dubius*) produce a leaf characterized as succulent and spinach-like. They are extremely popular around their regions of diversity, including India, and tropical South or Southeast Asia. A tremendous amount of diversity exists in these species and the wild, related species are sometimes used and or cultivated. The Organic Gardening and Farming Research Center of Rodale Press, Kutztown, PA began research on grain amaranths in 1975 and has been given responsibility for the characterization and increase of material distributed throughout the IBPGR global network.

Vegetable Crops. Root and tuber germplasm is, in general, not well represented in germplasm collections, with the exception of the potato. For the potato, most of IBPGR's work has been directed to providing assistance to national centers to collect and conserve local cultivated and wild relative germplasm. The primary responsibility for this crop lies with the Centro International de la Papa (CIP). The IBPGR has worked closely with CIAT and national programmes to collect and duplicate cultivated and wild cassava (Manihot sp.) germplasm. Extensive support has also been provided to CIAT to develop techniques for maintaining cassava germplasm in vitro and using cryopreservation techniques.

In addition to the collection and conservation of Ipomoea spp. germplasm in South America, Southeast Asia and the Pacific, IBPGR supported projects have included preliminary evaluation of morphological and cytological traits, in vitro conservation, and training in various aspects of sweet potato genetic resources. The collection of more regionally important roots and tubers have been accomplished through multi-crop or crop-specific missions. Representative genera include Colocasia, Dioscorea, Alocasia, and Xanthosoma.

There are several genera of root crops of Andean origin that have been collected and conserved througf, IBPGR support because of their importance in their native regions. *Oxalis tuberosa* and *Ullucus tuberosus* are of significant importance as starchy foods. The importance of *U. tuberosus* is often greater than that of *Oxalis*. Regional production is of such an extent that distribution to larger cities is possible. *Oxalis* and *Ullucus* ha ve potential in the U.S. as new crops because of their diversity in taste (ranging from sugary to acid) and tolerance of harsh growing conditions, including poor soils and high altitudes. Both of these crops, however, require additional research with regard to virus eradication. *Arracacia xanthorrhiza*, is a herbaceous perennial that produces smooth skinned starchy roots and edible stems that resemble celery. The roots produce a starch that is easily digested and has a delicate flavor and crisp texture, and is used to enhance stews and soups. Additional research on these three crops could reveal as yet unseen potential with regard to fillers and flavorings.

The IBPGR's work with the conservation of Latin American, African, and Asian vegetables began in 1980 and coincided with the push for conservation by national and regional programmes. This concurrent effort resulted in a doubling of collection sizes in five years. The IBPGR has devoted the greatest effort to the collection of landraces and primitive cultivars of *Lycopersicon, Capsicum, Allium, Abelmoschus, Brassica, Solanum* (eggplant), and several Cucurbitaceae genera. The collection of wild species for manv of these crops as well as multiplication and characterization of several collections has been partly or completely sponsored by IBPGR. The regional importance of many of these vegetables, justified the funding for conservation facilities in many countries in Africa, Asia and South America.

There are several vegetables of importance on a more local basis that have been conserved with IBPGR support. Cucurbits like *Sechium, Luffa,* and *Momordica* are currently underexploited in the U.S. In the Central American (*Sechium*) and sub-Asian (*Luffa* and *Momordica*) areas of origin (Smith 1976), the fruits from several species are used as important additions to the more common vegetables. The large, one-seeded fruits of *Sechium edule* are eaten raw or boiled and the roots are used as a starchy food. *Luffa cylindrica* and *L. acutangula* fruits are cooked, but dried fruits from *Luffa* are used as sponges.

The fruits of *Momordica charantia* are also used as cooked vegetables, but can also be pickled and used in curries. All of these are potentially valuable in extending the variety of vegetables that are currently in demand in oriental and vegetable dishes. Two *Brassicas, B. campestris* subsp. *chinensis* (pak-choi or Chinese mustard) and *B. campestris* subsp. *pekinensis* (Chinese cabbage) originated in China and are of major importance in East and Southeast Asia. The former produces shoots and the later heads of leaves. These species are now filling a void for domestic, low cost oriental vegetables. Both are reported to exist in hybrid varieties in Japan. Domestic ptoduction can be enhanced through the utilization of the variation that exists in several important germplasm collections.

Industrial Crops. Although the IBPGR works primarily with staple food crops, a number of industrial crops have also been incorporated when they are of importance in rural development. Most of the work to date has been with the genetic resources of the common crops such as beets, sugarcane, cacao, coconut, coffee, tea, cotton, rubber and oil palm. Principle emphasis with regard to collection was put on the well-known species of these crops, but some related and wild species were also collected. Other types of projects have also been supported including in vitro research on cacao.

Palms, representing several genera, have potential not only as sources of new U.S. crops, but also for enhancing the current germplasm base of the existing ornamental palms. Hearts of palm are found in the world market as exotic vegetables and salad additions. Hearts of palm can be produced in plantations from species of *Bactris, Euterpe, Sabal,* and *Cocos*. Edible palm oils can be derived from species in several palm genera including *Elaeis* and

Cocos. The IBPGR has sponsored collection and conservation of some of these genera.

Plant Breeding and Genetic Enhancement

Plant breeding as an art probably pre-dates civilization; it certainly has shown results (cultivated plant varieties) for the past 10,000 years. As a science and a full-time profession, however, plant breeding is largely a 20th century discipline. Its successes, as we all have heard many times over, have given rise to great expanses of a few highly favoured cultivars, which in turn have attracted epidemic spread of diseases and pests, uniquely adapted to their ever restricted range of hosts. From experiences with such epidemics, the concept of genetic vulnerability has been developed, along with its corollary, genetic diversity.

Promotion of genetic diversity requires introduction of new, unrelated breeding materials into basic and advanced breeding pools. Resulting new cultivars thus can display a greater diversity of genotypes, and present less opportunity, as grown on the farm, for epidemic disaster. But to bring greater diversity into the breeding pools means going outside the range of elite adapted local materials. It requires bringing in unadapted and, thus, unproductive cultivars—or even weedy or wild species—that have only a few useful traits, usually tightly linked with a panoply of other undesirable, yield-reducing traits. Such exotic material, when crossed directly with elite adapted lines, uniformly gives unsuccessful breeding results. Useful cultivars rarely or never can be developed from such wide crosses.

But skillful breeders can overcome this problem. They do not try, immediately, to use the unadapted material for production of elite new cultivars. They settle, instead, for selection of intermediates: lines or populations that, while maintaining the good new traits from the exotics, also contain some of the germplasm of the adapted elite cultivars. The broadening alien germplasm thus is at least partly naturalized, and most importantly, the new product, genetically enhanced breeding material, now is somewhat attractive to those plant breeders who are charged primarily with the task of reliably and repeatedly turning out top quality new cultivars.

Thus, the breeder who has performed the important function of partially acclimatizing alien germplasm, while conserving its essential genetic contributions, has performed the operation of "genetic enhancement." Genetic enhancement is needed to prevent genetic vulnerability brought about by the successes of modern plant breeding. Why is genetic enhancement thought of as a new requirement in plant breeding? Since all crop cultivars came originally from wild species, genetic enhancement—selection toward useful cultivar types—surely was needed for the original modification of the wild species.

There are at least three distinct uses of genetic enhancement. The first is to prevent genetic uniformity and consequent genetic vulnerability. Only recently has pre-breeding—genetic enhancement—become a necessary, frequent, and planned part of all plant breeding activities, an essential part of germplasm diversification strategies. Genetic enhancement has a second important purpose, that of raising yield levels to new heights. This goal is more often hoped for than achieved, but it is true that most break-through cultivars have highly diverse parentage. The semi-dwarf wheats, the high yield dwarf rices, the first U.S. hybrid sorghums, and even the first U.S. corn belt dent maize cultivars are examples.

In each case, extensive prebreeding preceded development of the breakthrough, high-yield cultivars. The prebreeding was used to adapt diverse kinds of germplasm to new genetic backgrounds and new geographic locales. A third use of genetic enhancement also needs mention. Genetic enhancement is used to bring in new quality traits not found in local cultivars. New levels of protein percentage in wheat or unusual starch properties in maize are examples.

Genetic enhancement is now necessary and useful-but it is not yet well-recognized as being so. This is evident because there is no group of scientists or professionals who call themselves "genetic enhancers", or "prebreeders"; there is no "American Society of Generic Enhancers," no "National Council of Commercial Pre-Breeders." Nor is there a recognized body of prebreeding theory, nor textbooks on the subject. Yet the job gets done-exotic germplasm is incorporated into adapted, elite stocks, and new cultivars often do contain significant segments of exotic germplasm, giving new kinds of disease and insect resistance, new levels or kinds of stress tolerance, or greater yield capability.

By and large, the job of generic enhancement is done by a special subset of plant breeders and geneticists in the public sector, scientists who either know they must pre-breed in order to make further progress in variety development of their crop, or know they like the challenge of finding, incorporating, and adapting useful germplasm from unlikely places. But in most cases, the main basis of their professional reputation, and of their ability to attract research funds, has been their successful production of finished cultivars. These scientists have supported their avocation, prebreeding, with their vocation, cultivar production.

An important deviation from this rule, that prebreeders also are cultivar developers, may be arising in the international research centers such as the International Rice Research Institute (IRRI), the International Crops Research Institute for the Semi-Arid Tropics (ICRISAT) or the International Maize and Wheat Improvement Center (CIMMYT). Such centers were formed with the express purpose of making cultivars and advanced breeding materials for use in developing countries. Today, as national breeding programmes are maturing in most developing countries, the centers are rethinking their missions.

Some people believe the centers should reduce or even cease cultivar development, in favour of development solely of genetically enhanced breeding materials, to be made available to all plant breeders in developing countries. Further, the centers often are sites for large germplasm depositories, and some believe the centers thus uniquely are suited to characterize and choose potentially useful materials in their collections of landraces and/or wild related species, and then to prebreed—genetically enhance—them for use by others for cultivar development.

These centers have the same problem, potentially, as breeders in U.S. land grant universities and other tax-supported institutions: in the absence of widespread recognition of genetic enhancement as a profession with easily identifiable products whose value can be calculated, centers that concentrate on genetic enhancement might lose their popular recognition and thus eventually the public funds that support them. This points up further the need for better definition and recognition of the products of genetic enhancement, and of its proper place in plant breeding and variety development. Cultivar development, of course, has had no lack of professional recognition; in fact, as a profession, it has become a successful commercial activity.

Many plant breeders now work in private industry, they are supported by the products of their activity-improved cultivars and hybrids-and their numbers increase, annually. But, plant breeders working for private industry never have done much prebreeding. Private breeders usually depend on the public sector to incorporate needed exotic genes, to remold exotic strains into adapted breeding materials. Genetically enhanced stocks, made by public breeders for their own use in cultivar production, have been freely available to private breeders as well, to use as they wish; privately employed breeders, pressed for short-term results, have depended heavily on publicly employed breeders for genetically enhanced stocks.

Trends are for public breeders to abandon cultivar production and move to more basic biological studies. Beginning young plant breeding scientists in the public sector are going in this more basic direction. Breeders of the older generation are not redirecting their activities, but they are retiring. The gradual change resulting from these two trends is breaking the connection in the chain of development from basic biology through genetic enhancement and on to cultivar production. Publicly employed plant biologists by and large are entering the plant breeding research field at levels well before genetic enhancement-in the more precisely defined (and better funded) science of molecular biology, for example. A gap thus may be developing, in the progression from basic plant biology to final cultivar production.

Some public breeders, seeing this problem, are trying to build up support specifically for genetic enhancement, but because genetic enhancement is not a well-recognized discipline—and perhaps because some people feel the term "genetic enhancement" is only a plant breeder's euphemism for final cultivar production—its support is neither large nor widespread. Private breeders, in general, have not taken up the torch of pre-breeding—or if so, are not talking about it. We may be building towards a shortage of genetically-enhanced breeding materials, a shortage that will become evident a few years further on.

Development of New Concept

Genetic enhancement, up to now, has utilized standard hybridization, segregation, and whole plant selection techniques. Backcrossing, cyclic population improvement, and pedigree selection among selfed progeny are examples of methods employed. But with the advent of molecular genetics and cell biology, a new kind of biotechnology-assisted genetic enhancement (or prebreeding) is possible. Heretofore unavailable genes for insect resistance may be transferred from alien species into elite genotypes (e.g., *Bt* genes from bacteria to maize); or, cellular fusions may be used to construct new systems of cytoplasmic male sterility (e.g., in *Brassica* species); or, restriction fragment length polymorphisms (RFLPs) may be used to build up tables of RFLP linkages with agronomically useful traits (e.g., as in wheat, to allow faster, planned introgression of alien germplasm). Such novel genetic potentials have raised the possibility of entirely new product lines-new kinds of plant cultivars, never before available.

The prospect of being able to market novel breeding stocks or novel varieties, produced via genetic engineering or other new cell and/or DNA technologies, has attracted the attention of entrepreneurially inclined scientists and business people. New companies, or new divisions in existing companies, have been formed to exploit the new commercial opportunities in plant biology. Commercial applications of biotechnology in plant prebreeding activities generally have been along one of two lines:

1. Some companies have developed fully integrated research and development programmes, designed to utilize a combination of molecular biology and standard plant breeding techniques for prebreeding and final cultivar production. Their intention is to turn out, and market, finished cultivars with novel traits or with greatly improved existing traits.
2. Other companies (or sometimes the same companies) have opted to utilize molecular genetics (for example, genetic transformation) or cell biology (for example, selection in tissue culture) to modify—to genetically enhance—existing elite germplasm stocks which then can be marketed as source materials for development of new cultivars containing the novel, engineered traits. The "engineered genetic enhancement" may be done on contract, or it may be done with expectation that the new enhanced genotypes will be marketed (via sale or licensing) to end-users such as seed companies or integrated food-processors. In either case, genetic enhancement is done as a commercial end in itself; the end-users, the purchasers of the enhanced materials, will do the final step of cultivar development.

Development of New Funding Base

But these biotechnology-using genetic enhancers are different from the earlier generation of genetic enhancers. Not only do they use different technology (and, often, different source materials, such as bacteria), they also have a different source of funding. They do not depend on public funds; they intend to be supported by profits from sale or licensing of their enhanced germplasm. They intend to be paid by the seed companies that use their products. The seed companies in turn are expected to raise the price of their seed as sold to the farmer, to cover the extra costs of cultivar development and thus keep profit margins at safe levels. So in the end, the farmer would be expected to pay for genetic enhancement, whereas up to now the public at large has borne the cost.

In some cases, genetically-enhanced materials, produced via new biotechnologies, might be used to reduce a seed company's costs of breeding or production, e.g., if new kinds of male sterile systems for hybrid seed production were to be developed. In such cases price increases for seed would not be required. But in general, genetic engineering is envisaged as adding a desired and affordable extra value—extra profitability—to crops grown by the farmer. The farmer would-according to this plan-be able to afford an extra charge for the seed because of the extra value in the seed.

The new payment scheme for generic enhancement thus opens the door for private enterprise and in theory will bring in funds needed for development of genetic enhancement. The scheme as yet exists only in theory, however. Up to this time, biotechnology-based genetic enhancement businesses have been supported mainly by venture capital funds, not by sale of genetically-enhanced product. Prices, and terms of sale, license or partnership only now are being worked out. And such arrangements necessarily must be made with no knowledge or experience of how much added value really will accrue from the genetically-enhanced materials, or when such value might be ready for sale in products, or whether farmers would want it, or be willing—and able—to pay for it. Thus, the new kind of genetic enhancement-genetic engineering enhancement-still is an unproven accomplishment, financially as well as technically.

If it succeeds, it will add a new dimension to the economics of farming, for farmers will be paying a larger share of the development cost of their seed inputs, and the general public (through general tax funds) will pay less. And a new category of food-and-agriculture-related workers in private enterprise will be formed: genetic engineering prebreeders. Such a development might erase the currently popular but really artificial distinction between biotechnology and older technologies, used in aid of genetic enhancement. Genetic enhancement's financial future is uncertain, as is its technological future.

Much will depend on whether some of its products obtained via any technology can be produced in relatively short term, with obvious and quantifiable value, and at a price that buyers can afford. If so, commercial funds and talent will be attracted; if not, commercial efforts at prebreeding as an end in itself will be abandoned. Then public funding, and scientists at public institutions, will need to continue to carry the load, unless the seed industry steps up rates and amounts of in-house activity in genetic enhancement, with or without aid of biotechnology.

Even if genetic enhancement does become a commercially successful enterprise, publicly-funded efforts will continue to be needed for those basic, preliminary kinds of genetic enhancement (early prebreeding) that require long-term effort with uncertain or even unpredictable results. Such long-term, chancy work will not attract private enterprise. And, many seed crops are not dealt with at all by private enterprise breeding, due to their small scale or otherwise low profit margin potential. Thus, for well-rounded genetic enhancement programmes in any crop, public research in genetic enhancement always will be needed.

Use of Biotechnology

Biotechnology-the sum of the technologies deriving from molecular and cell biology-will have utility in plant breeding beyond the prebreeding stages. Genetic transformation—nonsexual insertion of alien genes and/or gene regulating systems—is envisaged as eventually being applied, quickly and easily, to finished cultivars, to endow them with a series of useful new traits. However, experience is beginning to show that genetic background affects expression of genes and gene systems that are transferred in vitro, as is also the case when they are transferred via regular sexual mechanisms.

Long periods of trial and error are needed to bring transformed lines into desired levels of expression. Thus, for some years to come, genetic transformation perhaps will be used only as a prebreeding technique, one in which alien genes are put into breeding pools of elite germplasm, following which standard plant breeding methods are used to extract good new varieties containing the alien genes. Cell culture techniques—particularly protoplast fusion techniques—will have utility for making new combinations of nuclear and cytoplasmic genes. Immediate utility is seen for manufacturing new male sterility systems to make F_1 hybrid cultivars, or to develop new kinds of herbicide resistance.

With time and experience it may be possible to increase primary productive potential as well, by selecting and fixing optimum mitochondrial and chloroplast genotypes, balanced with proper nuclear genotypes. Such success seems far in the future, however, since at present little is known about optimization of nuclear, mitochondrial, and chloroplastic combinations. DNA technology will help plant breeding in a more traditional way, when information is further developed on maps of restriction fragment length polymorphisms (RFLPs) and their linkages with useful traits in crop plants.

In vitro tests, using RFLP markers, will allow near perfect accuracy in selecting for such useful linked traits. RFLP technology also will be used for precisely characterizing cultivar lineages and relationships, for positive identification of miss-identified or mislabeled cultivars, for characterizing gene frequencies of populations, and even to test hypotheses of landrace evolution, and global movement of crop genotypes.

Perhaps the greatest contribution of molecular biology to plant breeding will be a deeper and more precise understanding of gene action and inheritance, of cellular physiology and biochemistry, and perhaps of whole plant physiology. With such deepened understanding, breeding and selection via standard sexual methods will be planned and executed with greater vision and economy. Such knowledge and execution will be, in part, a consequence of efforts to make genetic transform-nation a practical reality. The fall-out of knowledge from genetic engineering efforts, wisely utilized by observant biologists, will be of great benefit to plant breeding, and to plant culture in general.

In addition to broadening the genetic base of established crops, genetic enhancement-and plant breeding in the sense of final cultivar development-can be used in two other ways: to develop new crops from heretofore uncultivated species, and to change old crops into new crops. Breeding of the first type (creating new crops where none existed before) usually is interpreted also to include development of advanced cultivars from landraces. Usually the advanced cultivars are made for industrial countries and the landraces come from developing countries. Soybeans and amaranth thus are spoken of as "new crops" in the United States, even though they have been cultivated for centuries as landraces in China, and in Central and South America, respectively

Challenges are great and are about the same, for changing either wild species or landraces into advanced cultivars, able to withstand monocropping and intensive culture at high production levels. Selection for productivity inevitably requires narrowing the germplasm base, in the course of selecting plant architecture and physiology able to withstand crowding and the stress of high yields, and amenable to machine planting, weeding and harvesting. Planting the narrowly selected genotypes in large areas gives excellent access to specially adapted disease and insect pests, as already noted, and so the need for tolerance or resistance always is next to be added to the list of breeding priorities for new crops.

Thus, genetic enhancement first narrows and then later must broaden the genetic base, in the course of new crop development. And as soon as users of the new product get used to larger quantities of it, they will want to standardize and then to change the quality of the product-as, in oil percentage, grain storability and hardness, or flavor and texture. The rule is simple: the more successful the crop, economically, the greater the demands on (and complaints about) the plant breeders.

Breeding of the second type-changing old crops into new ones-has been considered as a broad-scale possibility only since concepts of genetic engineering were developed. For example, ability to produce human insulin in bacteria gave rise to thoughts of producing insulin, or other useful primary gene products, in crop plants. Plants were envisaged as now able to turn sunshine, air and water (and soil nutrients) into high value specialty chemicals, instead of commodity chemicals like corn starch or soybean oil. Most of these ideas still are only ideas, but workers are proceeding to put some of the least complicated ones into action.

Modification, via genetic engineering, of the ratios of fatty acid components of rapeseed oil is contemplated, for example. Standard breeding to modify chemical components has been used in Canada already, to make rapeseed into a new crop: "canola." Canola oil, unlike standard rapeseed oil, is essentially free of erucic acid and glucosinolates and therefore is safe for human consumption when used as a cooking oil. Another possibility derives from recent success in cloning the R gene (first identified by Mendel) in peas which provides the opportunity to identify and transfer, in vitro, genes giving new starch branching properties and thus, potentially, useful new kinds of specialty starches in old starch-producing crops such as maize or wheat.

A word of caution is needed for those looking to replace major portions of commodity crop production areas with specialty-product new versions of the old crops. High value specialty products, by definition, require only small amounts of product to saturate the market. They won't replace very much of the production area now used for production of commodity crops. On the other hand, they will make intensive use of crop breeding and crop management skills. They are "brain-power intensive" crops, and as such, must command higher preunit prices—premiums—or they won't be grown.

Mass Selection

Mass selection is man's oldest plant breeding method. It was used 10,000 years ago to gradually change weedy grasses into the cereals—grain crops that today supply three-fourths of man's energy and over half of his protein. Mass selection is also one of the best plant breeding methods to enhance germplasm and develop improved cultivars today. To be effective, mass selection requires a variable, sexual, seed producing population of cross-pollinated plants. Genetic male-sterile genes can be used in self-pollinated species such as the soybean to create a cross-pollinated population. Enhancing such populations requires effective screens for the characters sought, effective methods for intermating the selected plants, and recurring cycles of these procedures to build up the frequency of the genes controlling the traits being selected.

Using mass selection to enhance traits desired in a good population avoids introducing undesirable traits that often come with exotic germplasm. Wayne Hanna, ARS geneticist at Tifton, Georgia, used *Pennisetum glaucum* var. *monodii* to supply dominant genes for immunity to rust, *Puccinia substriata* Ell. and Barth. var. *indica* Ramacha and Cumm. and leaf spot, *Pyricularia grisea* (Cke) Sarc. But crosses required to transfer the disease resistant genes also brought with them undesirable *monodii* genes for seed shattering, short-day sensitivity, sterile cytoplasm and susceptibility to *Helminthosporium* leaf spot.

Mass selection can retain in a population desirable characters not selected for or against. Increasing yield in Pensacola bahiagrass by mass selection did not alter its digestibility. Forages are better suited to mass selection enhancement than most crops. Many forage species are highly variable and uniformity is usually not required. Many are perennials and most important characters such as whole plant yield are multigenic in their inheritance. Seed yield is usually less important, characters can be selected visually, and cycles can often be shorter.

Mass selection has usually not been recommended as a breeding method to increase yield because heritability for spaced plant yields has usually been low. But heritability can be improved by keeping the environment uniform for all plants in a population from the time the seeds are planted until the spaced plants are selected. This has been a major objective as we have tried to improve mass selection as a plant breeding method to increase forage yields.

From 1960 to 1988, made a number of modifications and restrictions in conventional mass selection that have made it possible to produce one cycle per year with the same progress that made in the beginning when we used three years per cycle. The plant breeding method resulting from these modifications of mass selection we call 'recurrent restricted phenotypic selection (RRPS)'. RRPS in a closed-population of diploid, cross-pollinated Pensacola bahiagrass each year has effectively intermated the 200 largest visually-grid-selected plants from 1000 spaced plants, the largest seedlings in 20,000 RRPS intermated seedlings and has increased forage yield 16% per cycle through cycle 14 in spaced-plant-tests and 5% per cycle through cycle 9 in 3-year seeded-plot-tests.

Industrial Crops Breeding

The improvement of a new industrial crop is not appreciably different from enhancement and breeding of conventional food, feed or fiber crops. The major difference is that the plant breeder often starts with a different, and frequently unique or exotic germplasm base from which to develop a plant that produces significant yields of relatively high valued industrial feedstocks and/or products. These products may take various forms such as fibers, rubber, resins, waxes, gums, pharmaceuticals, biologically active materials, essential oils, and other oils with unique fatty acids.

The basic breeding and genetic methods utilized in the development of usable cultivars of a new industrial crop are essentially the same as for conventional crops. One differing aspect is that the breeder is often totally unfamiliar with the species, the germplasm, and potential end products. In most instances, this necessitates a very thorough study and literature review, and familiarization with all aspects of the breeding and developmental processes to a much greater extent than is usual when working with a conventional crop.

In many instances pertinent literature may not be available except for some obscure taxonomic treatise or reference to an analysis in some chemical journal. Interestingly, some good information is frequently unearthed from obscure, often foreign journals. One plus for the new crop researcher is that almost anything you discover is of interest and may well merit publication—a real advantage in these times of performance standards, tenure, and the paper chase.

The total system of new crops research and development is composed of five major components or activities. The steps proceed in an essentially sequential fashion, but they are not discrete, and varying amounts of overlap may occur among them. The first three phases involve three major activities of the germplasm system—collection, evaluation, and enhancement and development. The fourth major activity involves the utilization of germplasm in cultivar development, which is intimately involved with agronomic/horticultural evaluation, and the development of appropriate cultural and management systems.

The last, but not least, involves an array of activities associated with full scale commercialization. Types of activities and inputs of the various scientific disciplines and industrial components are indicated within each of the five areas of the total system. It is clear that the plant breeder/geneticist is a key figure in the developmental process. New crop breeders, because of their unique role in the total process, frequently must provide the overall coordination, and manage and effectively utilize all available resources and appropriate methods to see that the developmental process moves forward.

Each new crop is likely to have its own, unique developmental pattern. The amount of time and resources required and devoted to the various phases will vary depending upon numerous factors. Three major factors that influence the timing and pattern of development of a new crop is whether it is developed by (1) domesticating a wild species, (2) adapting an existing crop from another area, or (3) making genetic changes in an established crop so that a new commodity is produced. In addition, there are other important factors that determine the strategy and methods employed in the developmental process. These include the availability nature, and adaptation of germplasm resources for the species; the extent of previously conducted research and development on the targeted crop species; the existence of usable and appropriate technology for evaluation, selection, and breeding for the products and coproducts sought; the need and urgency for completion of the developmental process; and the potential economic impact, which is the ultimate driving force of the process.

A plant breeder must be flexible in the approach to breeding a new crop where much is unknown. The breeder must be innovative and able to shift gears rapidly to meet the opportunities and constraints as they are encountered.

9

Experiments in Plant Hybridisation

Mendel's this experiment was practically confined to a small plant group, and is now, after eight years' pursuit, concluded in all essentials. Whether the plan upon which the separate experiments were conducted and carried out was the best suited to attain the desired end is left to the friendly decision of the reader. The value and utility of any experiment are determined by the fitness of the material to the purpose for which it is used, and thus in the case before us it cannot be immaterial what plants are subjected to experiment and in what manner such experiment is conducted.

The selection of the plant group which shall serve for experiments of this kind must be made with all possible care if it be desired to avoid from the outset every risk of questionable results. The experimental plants must necessarily:

— Possess constant differentiating characteristics

— The hybrids of such plants must, during the flowering period, be protected from the influence of all foreign pollen, or be easily capable of such protection.

— The hybrids and their offspring should suffer no marked disturbance in their fertility in the successive generations.

Accidental impregnation by foreign pollen, if it occurred during the experiments and were not recognized, would lead to entirely erroneous conclusions. Reduced fertility or entire sterility of certain forms, such as occurs in the offspring of many hybrids, would render the experiments very difficult or entirely frustrate them. In order to discover the relations in which the hybrid forms stand towards each other and also towards their progenitors it appears to be necessary that all member of the series developed in each successive generations should be, *without exception*, subjected to observation.

At the very outset special attention was devoted to the *Leguminosae* on account of their peculiar floral structure. Experiments which were made with several members of this family led to the result that the genus *Pisum* was found to possess the necessary qualifications. Some thoroughly distinct forms of this genus possess characters which are constant, and easily and certainly recognizable, and when their hybrids are mutually crossed they yield perfectly fertile progeny. Furthermore, a disturbance through foreign pollen cannot easily occur, since the fertilizing organs are closely packed inside the keel and the anthers burst within the bud, so that the stigma becomes covered with pollen even before the flower opens. This circumstance is especially important.

As additional advantages worth mentioning, there may be cited the easy culture of these plants in the open ground and in pots, and also their relatively short period of growth. Artificial fertilization is certainly a somewhat elaborate process, but nearly always succeeds. For this purpose the bud is opened before it is perfectly developed, the keel is removed, and each stamen carefully extracted by means of forceps, after which the stigma can at once be dusted over with the foreign pollen. In all, 34 more or less distinct varieties of Peas were obtained from several seedsmen and subjected to a two year's trial. In the case of one variety there were noticed, among a larger number of plants all alike, a few forms which were markedly different.

These, however, did not vary in the following year, and agreed entirely with another variety obtained from the same seedsman; the seeds were therefore doubtless merely accidentally mixed. All the other varieties yielded perfectly constant and similar offspring; at any rate, no essential difference was observed during two trial years. For fertilization 22 of these were selected and cultivated during the whole period of the experiments. They remained constant without any exception.

Their systematic classification is difficult and uncertain. If we adopt the strictest definition of a species, according to which only those individuals belong to a species which under precisely the same circumstances display precisely similar characters, no two of these varieties could be referred to one species. According to the opinion of experts, however, the majority belong to the species Pisum sativum; while the rest are regarded and classed, some as sub-species of P. sativum, and some as independent species, such as P. quadratum, P. saccharatum, and P. umbellatum. The positions, however, which may be assigned to them in a classificatory system are quite immaterial for the purposes of the experiments in question. It has so far been found to be just as impossible to draw a sharp line between the hybrids of species and varieties as between species and varieties themselves.

Division and Arrangement

If two plants which differ constantly in one or several characters be crossed, numerous experiments have demonstrated that the common characters are transmitted unchanged to the hybrids and their progeny; but each pair of differentiating characters, on the other hand, unite in the hybrid to form a new character, which in the progeny of the hybrid is usually variable. The object of the experiment was to observe these variations in the case of each pair of differentiating characters, and to deduce the law according to which they appear in successive generations. The experiment resolves itself therefore into just as many separate experiments are there are constantly differentiating characters presented in the experimental plants.

The various forms of Peas selected for crossing showed differences in length and color of the stem; in the size and form of the leaves; in the position, color, size of the flowers; in the length of the flower stalk; in the color, form, and size of the pods; in the form and size of the seeds; and in the color of the seed-coats and of the albumen(endosperm). Some of the characters noted do not permit of a sharp and certain separation, since the difference is of a "more or less" nature, which is often difficult to define. Such characters could not be utilized for the separate experiments; these could only be applied to characters which stand out clearly and definitely in the plants. Lastly, the result must show whether they, in their entirety, observe a regular behavior in their hybrid unions, and whether from these facts any conclusion can be reached regarding those characters which possess a subordinate significance in the type.

The characters which were selected for experiment relate:

— To the *difference in the form of the ripe seeds.* These are either round or roundish, the depressions, if any, occur on the surface, being always only shallow; or they are irregularly angular and deeply wrinkled (P. quadratum).

— To the *difference in the color of the seed albumen* (endosperm). The albumen of the ripe seeds is either pale yellow, bright yellow and orange colored, or it possesses a more or less intense green tint. This difference of color is easily seen in the seeds as their coats are transparent.

— To the *difference in the color of the seed-coat.* This is either white, with which character white flowers are constantly correlated; or it is gray, gray-brown, leather-brown, with or without violet spotting, in which case the color of the standards is violet, that of the wings purple, and the stem in the axils of the leaves is of a reddish tint. The gray seed-coats become dark brown in boiling water.

— To the *difference in the form of the ripe pods.* These are either simply inflated, not contracted in places; or they are deeply constricted between the seeds and more or less wrinkled (P. saccharatum).

— To the *difference in the color of the unripe pods.* They are either light to dark green, or vividly yellow, in which coloring the stalks, leaf-veins, and calyx participate.

— To the *difference in the position of the flowers.* They are either axial, that is, distributed along the main stem; or they are terminal, that is, bunched at the top of the stem and arranged almost in a false umbel; in this case the upper part of the stem is more or less widened in section (P. umbellatum).

— To the *difference in the length of the stem.* The length of the stem is very various in some forms; it is, however, a constant character for each, in so far that healthy plants, grown in the same soil, are only subject to unimportant variations in this character. In experiments with this character, in order to be able to discriminate with certainty, the long axis of 6 to 7 ft. was always crossed with the short one of 3/4 ft. to 1 [and] 1/2 ft.

Each two of the differentiating characters enumerated above were united by cross-fertilization. There were made for the

1st experiment 60 fertilizations on 15 plants.
2nd experiment 58 fertilizations on 10 plants.
3rd experiment 35 fertilizations on 10 plants.
4th experiment 40 fertilizations on 10 plants.
5th experiment 23 fertilizations on 5 plants.
6th experiment 34 fertilizations on 10 plants.
7th experiment 37 fertilizations on 10 plants.

From a larger number of plants of the same variety only the most vigorous were chosen for fertilization. Weakly plants always afford uncertain results, because even in the first generation of hybrids, and still more so in the subsequent ones, many of the offspring either entirely fail to flower or only form a few and inferior seeds. Furthermore, in all the experiments reciprocal crossings were effected in such a way that each of the two varieties which in one set of fertilizations served as seed-bearer in the other set was used as the pollen plant.

The plants were grown in garden beds, a few also in pots, and were maintained in their natural upright position by means of sticks, branches of trees, and strings stretched between. For each experiment a number of pot plants were placed during the blooming period in a greenhouse, to serve as control plants for the main experiment in the open as regards possible disturbance by insects. Among the insects which visit Peas the beetle *Bruchus pisi* might be detrimental to the experiments should it appear in numbers.

The female of this species is known to lay the eggs in the flower, and in so doing opens the keel; upon the tarsi of one specimen, which was caught in a flower, some pollen grains could clearly be seen under a lens. Mention must also be made of a circumstance which possibly might lead to the introduction of foreign pollen. It occurs, for instance, in some rare cases that certain parts of an otherwise normally developed flower wither, resulting in a partial exposure of the fertilizing organs. A defective development of the keel has also been observed, owing to which the stigma and anthers remained partially covered. It also sometimes happens that the pollen does not reach full perfection. In this event there occurs a gradual lengthening of the pistil during the blooming period, until the stigmatic tip protrudes at the point of the keel. This remarkable appearance has also been observed in hybrids of Phaseolus and Lathyrus.

The risk of false impregnation by foreign pollen is, however, a very slight one with Pisum, and is quite incapable of disturbing the general result. Among more than 10,000 plants which were carefully examined there were only a very few cases where an indubitable false impregnation had occurred. Since in the greenhouse such a case was never remarked, it may well be supposed that *Bruchus pisi*, and possibly also the described abnormalities in the floral structure, were to blame.

Forms of the Hybrids

Experiments which in previous years were made with ornamental plants have already affording evidence that the hybrids, as a rule, are not exactly intermediate between the parental species. With some of the more striking characters, those, for instance, which relate to the form and size of the leaves, the pubescence of the several parts, etc., the intermediate, indeed, is nearly always to be seen; in other cases, however, one of the two parental characters is so preponderant that it is difficult, or quite impossible, to detect the other in the hybrid.

This is precisely the case with the Pea hybrids. In the case of each of the 7 crosses the hybrid-character resembles that of one of the parental forms so closely that the other either escapes observation completely or cannot be detected with certainty. This circumstance is of great importance in the determination and classification of the forms under which the offspring of the hybrids appear. Henceforth in this paper those characters which are transmitted entire, or almost unchanged in the hybridization, and therefore in themselves constitute the characters of the hybrid, are termed the *dominant,* and those which become latent in the process *recessive*. The expression "recessive" has been chosen because the characters thereby designated withdraw or entirely disappear in the hybrids, but nevertheless reappear unchanged in their progeny, as will be demonstrated later on.

It was furthermore shown by the whole of the experiments that it is perfectly immaterial whether the dominant character belongs to the seed plant or to the pollen plant; the form of the hybrid remains identical in both cases. This interesting fact was also emphasized by Gärtner, with the remark that even the most practiced expert is not in a position to determine in a hybrid which of the two parental species was the seed or the pollen plant.

Of the differentiating characters which were used in the experiments the following are dominant:

- — The round or roundish form of the seed with or without shallow depressions.
- — The yellow coloring of the seed albumen.
- — The gray, gray-brown, or leather brown color of the seed-coat, in association with violet-red blossoms and reddish spots in the leaf axils.
- — The simply inflated form of the pod.
- — The green coloring of the unripe pod in association with the same color of the stems, the leaf-veins and the calyx.
- — The distribution of the flowers along the stem.
- — The greater length of stem.

With regard to this last character it must be stated that the longer of the two parental stems is usually exceeded by the hybrid, a fact which is possibly only attributable to the greater luxuriance which appears in all parts of plants when stems of very different lengths are crossed. Thus, for instance, in repeated experiments, stems of 1 ft. and 6 ft. in length yielded without exception hybrids which varied in length between 6 ft. and 7 [and] 1/2 ft.

The *hybrid seeds* in the experiments with seed-coat are often more spotted, and the spots sometimes coalesce into small bluish-violet patches. The spotting also frequently appears even when it is absent as a parental character. The hybrid forms of the *seed-shape* and of the *[color of the] albumen* are developed immediately after the artificial fertilization by the mere influence of the foreign pollen. They can, therefore, be observed even in the first year of experiment, whilst all the other characters naturally only appear in the following year in such plants as have been raised from the crossed seed.

First Generation From the Hybrids

In this generation there reappear, *together with the dominant* characters, also the *recessive* ones with their peculiarities fully developed, and this occurs in the definitely expressed average proportion of 3:1, so that among each 4 plants of this generation 3 display the dominant character and one the recessive. This relates without exception to all the characters which were investigated in the experiments. The angular wrinkled form of the seed, the green color of the albumen, the white color of the seed-coats and the flowers, the constriction of the pods, the yellow color of the unripe pod, of the stalk, of the calyx, and of the leaf venation, the umbel-like form of the inflorescence, and the dwarfed stem, all reappear in the numerical proportion given, without any essential alteration. *Transitional forms were not observed in any experiment.*

Since the hybrids resulting from reciprocal crosses are formed alike and present no appreciable difference in their subsequent development, consequently these results can be reckoned together in each experiment. The relative numbers which were obtained for each pair of differentiating characters are as follows:

— Expt. 1: Form of seed. From 253 hybrids 7324 seeds were obtained in the second trial year. Among them were 5474 round or roundish ones and 1850 angular wrinkled ones. Therefrom the ratio 2.96:1 is deduced.

— Expt. 2: Color of albumen.. 258 plants yielded 8023 seeds, 6022 yellow, and 2001 green; their ratio, therefore, is as 3.01:1.

In these two experiments each pod yielded usually both kinds of seed. In well-developed pods which contained on the average 6 to 9 seeds, it often happened that all the seeds were round (Expt. 1) or all yellow (Expt. 2); on the other hand there were never observed more than 5 wrinkled or 5 green ones on one pod. It appears to make no difference whether the pods are developed early or later in the hybrid or whether they spring from the main axis or from a lateral one. In some few plants only a few seeds developed in the first formed pods, and these possessed exclusively one of the two characters, but in the subsequently developed pods the normal proportions were maintained nevertheless.

As in separate pods, so did the distribution of the characters vary in separate plants. By way of illustration the first 10 individuals from both series of experiments may serve.

As extremes in the distribution of the two seed characters in *one* plant, there were observed in Expt. 1 an instance of 43 round and only 2 angular, and another of 14 round and 15 angular seeds. In Expt. 2 there was a case of 32 yellow and only 1 green seed, but also one of 20 yellow and 19 green. These two experiments are important for the determination of the average ratios, because with a smaller number of experimental plants they show that very considerable fluctuations may occur. In counting the seeds, also, especially in Expt. 2, some care is requisite, since in some of the seeds of many plants the green color of the albumen is less developed, and at first may be easily overlooked.

The cause of this partial disappearance of the green coloring has no connection with the hybrid-character of the plants, as it likewise occurs in the parental variety. This peculiarity is also confined to the individual and is not inherited by the offspring. In luxuriant plants this appearance was frequently noted. Seeds which are damaged by insects during their development often vary in color and form, but with a little practice in sorting, errors are easily avoided. It is almost superfluous to mention that the pods must remain on the plants until they are thoroughly ripened and have become dried, since it is only then that the shape and color of the seed are fully developed.

— Expt. 3: Color of the seed-coats. Among 929 plants, 705 bore violet-red flowers and gray-brown seed-coats; 224 had white flowers and white seed-coats, giving the proportion 3.15:1.

— Expt. 4: Form of pods. Of 1181 plants, 882 had them simply inflated, and in 299 they were constricted. Resulting ratio, 2.95:1.

— Expt. 5: Color of the unripe pods. The number of trial plants was 580, of which 428 had green pods and 152 yellow ones. Consequently these stand in the ratio of 2.82:1.

— Expt. 6: Position of flowers. Among 858 cases 651 had inflorescences axial and 207 terminal. Ratio, 3.14:1.

— Expt. 7: Length of stem. Out of 1064 plants, in 787 cases the stem was long, and in 277 short. Hence a mutual ratio of 2.84:1. In this experiment the dwarfed plants were carefully lifted and transferred to a special bed. This precaution was necessary, as otherwise they would have perished through being overgrown by their tall relatives. Even in their quite young state they can be easily picked out by their compact growth and thick dark-green foliage.

If now the results of the whole of the experiments be brought together, there is found, as between the number of forms with the dominant and recessive characters, an average ratio of 2.98:1, or 3:1. The dominant character can have here a *double signification*; namely, that of either a parental character or a hybrid-character. In which of the two significations it appears in each separate case can only be decided in the following generation. As a parental character it must pass over unchanged to the whole of the offspring; as a hybrid-character, on the other hand, it must maintain the same behavior as in the first generation.

Second Generation From the Hybrids

Those forms which in the first generation exhibit the recessive character do not further vary in the second generation as regards this character; they remain *constant* in their offspring. It is otherwise with those which possess the dominant character in the first generation. Of these *two*-thirds yield offspring which display the dominant and recessive characters in the proportion of 3:1, and thereby show exactly the same ratio as the hybrid forms, while only *one*-third remains with the dominant character constant.

The separate experiments yielded the following results:

- Expt. 1: Among 565 plants which were raised from round seeds of the first generation, 193 yielded round seeds only, and remained therefore constant in this character; 372, however, gave both round and wrinkled seeds, in the proportion of 3:1. The number of the hybrids, therefore, as compared with the constants is 1.93:1.
- Expt. 2: Of 519 plants which were raised from seeds whose albumen was of yellow color in the first generation, 166 yielded exclusively yellow, while 353 yielded yellow and green seeds in the proportion of 3:1. There resulted, therefore, a division into hybrid and constant forms in the proportion of 2.13:1.

For each separate trial in the following experiments 100 plants were selected which displayed the dominant character in the first generation, and in order to ascertain the significance of this, ten seeds of each were cultivated.

— Expt. 3: The offspring of 36 plants yielded exclusively gray-brown seed-coats, while of the offspring of 64 plants some had gray-brown and some had white.

— Expt. 4: The offspring of 29 plants had only simply inflated pods; of the offspring of 71, on the other hand, some had inflated and some constricted.

— Expt. 5: The offspring of 40 plants had only green pods; of the offspring of 60 plants some had green, some yellow ones.

— Expt. 6.: The offspring of 33 plants had only · axial flowers; of the offspring of 67, on the other hand, some had axial and some terminal flowers.

— Expt. 7: The offspring of 28 plants inherited the long axis, of those of 72 plants some the long and some the short axis.

In each of these experiments a certain number of the plants came constant with the dominant character. For the determination of the proportion in which the separation of the forms with the constantly persistent character results, the two first experiments are especially important, since in these a larger number of plants can be compared. The ratios 1.93:1 and 2.13:1 gave together almost exactly the average ratio of 2:1. Experiment 6 gave a quite concordant result; in the others the ratio varies more or less, as was only to be expected in view of the smaller number of 100 trial plants. Experiment 5, which shows the greatest departure, was repeated, and then in lieu of the ratio of 60:40, that of 65:35 resulted. *The average ratio of 2:1 appears, therefore, as fixed with certainty.* It is therefore demonstrated that, of those forms which posses the dominant character in the first generation, two-thirds have the hybrid-character, while one-third remains constant with the dominant character.

The ratio 3:1, in accordance with which the distribution of the dominant and recessive characters results in the first generation, resolves itself therefore *in all experiments into the ratio of 2:1:1,* if the dominant character be differentiated according to its significance as a hybrid-character or as a parental one. Since the members of the first generation spring directly from the seed of the hybrids, it is now clear that the hybrids form seeds having one or other of the two differentiating characters, and of these one-half develop again the hybrid form, while the other half yield plants which remain constant and receive the dominant or the recessive characters in equal numbers.

Subsequent Generations From the Hybrids

The proportions in which the descendants of the hybrids develop and split up in the first and second generations presumably hold good for all subsequent progeny. Experiments 1 and 2 have already been carried through 6 generations; 3 and 7 through 5; and 4, 5, and 6 through 4; these experiments being continued from the third generation with a small number of plants, and no departure from the rule has been perceptible. The offspring of the hybrids separated in each generation in the ratio of 2:1:1 into hybrids and constant forms.

If **A** be taken as denoting one of the two constant characters, for instance the dominant, a the recessive, and Aa the hybrid form in which both are conjoined, the expression

$$\mathbf{A + 2Aa + a}$$

shows the terms in the series for the progeny of the hybrids of two differentiating characters.

The observation made by Gärtner, Kölreuter, and others, that hybrids are inclined to revert to the parental forms, is also confirmed by the experiments described. It is seen that the number of the hybrids which arise from one fertilization, as compared with the number of forms which become constant, and their progeny from generation to generation, is continually diminishing, but that nevertheless they could not entirely disappear.

If an average equality of fertility in all plants in all generations be assumed, and if, furthermore, each hybrid forms seed of which one-half yields hybrids again, while the other half is constant to both characters in equal proportions, the ratio of numbers for the offspring in each generation is seen by the following summary, in which A and a denote again the two parental characters, and Aa the hybrid forms. For brevity's sake it may be assumed that each plant in each generation furnishes only 4 seeds.

In the tenth generation, for instance, $2^n - 1 = 1023$. There result, therefore, in each 2048 plants which arise in this generation 1023 with the constant dominant character, 1023 with the recessive character, and only two hybrids.

Offspring of Hybrids in Differentiating Characters

The next task consisted in ascertaining whether the law of development discovered in these applied to each pair of differentiating characters when several diverse characters are united in the hybrid by crossing. As regards the form of the hybrids in these cases, the experiments showed throughout that this invariably more nearly approaches to that one of the two parental plants which possesses the greater number of dominant characters. If, for instance, the seed plant has a short stem, terminal white flowers, and simply inflated pods; the pollen plant, on the other hand, a long stem, violet-red flowers distributed along the stem, and constricted pods; the hybrid resembles the seed parent only in the form of the pod; in the other characters it agrees with the pollen parent. Should one of the two parental types possess only dominant characters, then the hybrid is scarcely or not at all distinguishable from it.

Two experiments were made with a considerable number of plants. In the first experiment the parental plants differed in the form of the seed and in the color of the albumen; in the second in the form of the seed, in the color of the albumen, and in the color of the seed-coats. Experiments with seed characters give the result in the simplest and most certain way.

In order to facilitate study of the data in these experiments, the different characters of the seed plant will be indicated by A, B, C, those of the pollen plant by a, b, c, and the hybrid forms of the characters by Aa, Bb, and Cc.

First Experiment: **AB** Seed parents, **abc** Pollen parents,

A form round **a** form wrinkled

B albumen yellow **b** albumen green

The fertilized seeds appeared round and yellow like those of the seed parents. The plants raised therefrom yielded seeds of four< sorts, which frequently presented themselves in one pod. In all, 556 seeds were yielded by 15 plants, and of these there were:

- 315 round and yellow,
- 101 wrinkled and yellow,
- 108 round and green,
- 32 wrinkled and green.

All were sown the following year. 11 of the round yellow seeds did not yield plants, and 3 plants did not form seeds. Among the rest:

38 had round yellow seeds .... AB
65 round yellow and green seeds..................ABb
60 round yellow and wrinkled yellow seeds...............AaB
138 round yellow and green, wrinkled yellow and green seeds.......... AaBb

From the wrinkled yellow seeds 96 resulting plants bore seed, of which:

— 28 had only wrinkled yellow seeds................aB

— 68 wrinkled yellow and green seeds ....aBb

From 108 round green seeds 102 resulting plants fruited, of which:

— 35 had only round green seeds ....Ab

— 67 round and wrinkled green seeds ....Aab

— The wrinkled green seeds yielded 30 plants which bore seeds all of like character; they remained constant ......ab.

The offspring of the hybrids appeared therefore under 9 different forms, some of them in very unequal numbers. When these are collected and coordinated we find:

38	plants with the sign			AB
35	"	"	"	Ab
28	"	"	"	aB
30	"	"	"	ab
65	"	"	"	ABb
68	"	"	"	aBb
60	"	"	"	AaB
67	"	"	"	Aab
138	"	"	"	AaBb

The whole of the forms may be classed into 3 essentially different groups. The first includes those with the signs AB, Ab, aB, and ab: they possess only constant characters and do not vary again in the next generation. Each of these forms is represented on the average 33 times. The second group includes the signs ABb, aBb, AaB, Aab: these are constant in one character and hybrid in another, and vary in the next generation only as regards the hybrid-character. Each of these appears on any average 65 times. The form AaBb occurs 138 times: it is hybrid in both characters, and behaves exactly as do the hybrids from which it is derived.

If the numbers in which the forms belonging to these classes appear be compared, the ratios of 1:2:4 are unmistakably evident. The numbers 33, 65, 138 present very fair approximations to the ratio numbers of 33, 66, 132.

The development series consists, therefore, of 9 classes, of which 4 appear therein always once and are constant in both characters; the forms AB, ab, resemble the parental forms, the two others present combinations between the conjoined characters A, a, B, b, which combinations are likewise possibly constant. Four classes appear always twice, and are constant in one character and hybrid in the other. One class appears four times, and is hybrid in both characters. Consequently, the offspring of the hybrids, if two kinds of differentiating characters are combined therein, are represented by the expression

AB + Ab + aB + ab + 2ABb + 2aBb + 2AaB + 2Aab + 4AaBb

This expression is indisputably a combination series in which the two expressions for the characters A and a, B and b are combined. We arrive at the full number of the classes of the series by the combination of the expressions:

A + 2Aa + a

B + 2Bb + b

Second Experiment:

ABC Seed parents,	abc Pollen parents,
A form round	a form wrinkled
B albumen yellow	b albumen green
C seed-coat gray-brown	c seed-coat white

This experiment was made in precisely the same way as the previous one. Among all the experiments it demanded the most time and trouble. From 24 hybrids 687 seeds were obtained in all: these were all either spotted, gray-brown or gray-green, round or wrinkled. From these in the following year 639 plants fruited, and as further investigation showed, there were among them:

8 plants ABC	22 plants ABCc	45 plants ABbCc
14 " ABc	17 " AbCc	36 " aBbCc
9 " AbC	25 " aBCc	38 " AaBCc
11 " Abc	20 " abCc	40 " AabCc
8 " aBC	15 " ABbC	49 " AaBbC
10 " aBc	18 " ABbc	48 " AaBbc
10 " abC	19 " aBbC	
7 " abc	24 " aBbc	
	14 " AaBC	78 " AaBbCc
	18 " AaBc	
	20 " AabC	
	16 " Aabc	

The whole expression contains 27 terms. Of these 8 are constant in all characters, and each appears on the average 10 times; 12 are constant in two characters, and hybrid in the third; each appears on the average 19 times; 6 are constant in one character and hybrid in the other two; each appears on the average 43 times. One form appears 78 times and is hybrid in all of the characters. The ratios 10:19:43:78 agree so closely with the ratios 10:20:40:80, or 1:2:4:8 that this last undoubtedly represents the true value.

The development of the hybrids when the original parents differ in 3 characters results therefore according to the following expression:

ABC + ABc + AbC + Abc + aBC + aBc + abC + abc +
2ABCc + 2AbCc + 2aBCc + 2abCc + 2ABbC + 2ABbc +
2aBbC + 2aBbc + 2AaBC + 2AaBc + 2AabC + 2Aabc +
4ABbCc + 4aBbCc + 4AaBCc + 4AabCc + 4AaBbC +
4AaBbc + 8AaBbCc.

Here also is involved a combination series in which the expressions for the characters A and a, B and b, C and c, are united. The expressions:

A + 2Aa + a
B + 2Bb + b
C + 2Cc + c

give all the classes of the series. The constant combinations which occur therein agree with all combinations which are possible between the characters A, B,C,a,b,c; two thereof, ABC and abc, resemble the two original parental stocks. In addition, further experiments were made with a smaller number of experimental plants in which the remaining characters by twos and threes were united as hybrids: all yielded approximately the same results. There is therefore no doubt that for the whole of the characters involved in the experiments the principle applies that the offspring of the hybrids in which several essentially different characters are combined exhibit the terms of a series of combinations, in which the developmental series for each pair of differentiating characters are united. It is demonstrated at the same time that the relation of each pair of different characters in hybrid union is independent of the other differences in the two original parental stocks.

If *n* represent the number of the differentiating characters in the two original stocks, 3^*n* gives the number of terms of the combination series, 4^*n* the number of individuals which belong to the series, and 2^n the number of unions which remain constant. The series therefore contains, if the original stocks differ in four characters, $3^4 = 81$ classes, $4^4 = 256$ individuals, and $2^4 = 16$ constant forms: or, which is the same, among each 256 offspring of the hybrids are 81 different combinations, 16 of which are constant.

All constant combinations which in Peas are possible by the combination of the said 7 differentiating characters were actually obtained by repeated crossing. Their number is given by $2^7 = 128$. Thereby is simultaneously given the practical proof that the constant characters which appear in the several varieties of a group of plants may be obtained in all the associations which are possible according to the laws of combination, by means of repeated artificial fertilization.

As regards the flowering time of the hybrids, the experiments are not yet concluded. It can, however, already be stated that the time stands almost exactly between those of the seed and pollen parents, and that the constitution of the hybrids with respect to this character probably follows the rule ascertained in the case of the other characters. The forms which are selected for experiments of this class must have a difference of at least 20 days from the middle flowering period of one to that of the other; furthermore, the seeds when sown must all be placed at the same depth in the earth, so that they may germinate simultaneously. Also, during the whole flowering period, the more important variations in temperature must be taken into account, and the partial hastening or delaying of the flowering which may result therefrom. It is clear that this experiment presents many difficulties to be overcome and necessitates great attention.

If we endeavor to collate in a brief form the results arrived at, we find that those differentiating characters, which admit of easy and certain recognition in the experimental plants, *all behave exactly alike in their hybrid associations*. The offspring of the hybrids of each pair of differentiating characters are, one-half, hybrid again, while the other half are constant in equal proportions having the characters of the seed and pollen parents respectively. If several differentiating characters are combined by cross-fertilization in a hybrid, the resulting offspring form the terms of a combination series in which the combination series for each pair of differentiating characters are united.

The uniformity of behavior shown by the whole of the characters submitted to experiment permits, and fully justifies, the acceptance of the principle that a similar relation exists in the other characters which appear less sharply defined in plants, and therefore could not be included in the separate experiments. An experiment with peduncles of different lengths gave on the whole a fairly satisfactory results, although the differentiation and serial arrangement of the forms could not be effected with that certainty which is indispensable for correct experiment.

Reproductive Cells of the Hybrids

The results of the previously described experiments led to further experiments, the results of which appear fitted to afford some conclusions as regards the composition of the egg and pollen cells of hybrids. An important clue is afforded in Pisum by the circumstance that among the progeny of the hybrids constant forms appear, and that this occurs, too, in respect of all combinations of the associated characters. So far as experience goes, we find it in every case confirmed that constant progeny can only be formed when the egg cells and the fertilizing pollen are of like character, so that both are provided with the material for creating quite similar individuals, as is the case with the normal fertilization of pure species.

We must therefore regard it as certain that exactly similar factors must be at work also in the production of the constant forms in the hybrid plants. Since the various constant forms are produced in *one* plant, or even in *one* flower of a plant, the conclusion appears logical that in the ovaries of the hybrids there are formed as many sorts of egg cells, and in the anthers as many sorts of pollen cells, as there are possible *constant* combination forms, and that these egg and pollen cells agree in their internal compositions with those of the separate forms.

In point of fact it is possible to demonstrate theoretically that this hypothesis would fully suffice to account for the development of the hybrids in the separate generations, if we might at the same time assume that the various kinds of egg and pollen cells were formed in the hybrids on the average in equal numbers.

In order to bring these assumptions to an experimental proof, the following experiments were designed. Two forms which were constantly different in the form of the seed and the color of the albumen were united by fertilization.

If the differentiating characters are again indicated as A, B, a, b, we have:

Second Experiment:

AB Seed parents,	ab Pollen parents,
A form round	a form wrinkled
B albumen yellow	b albumen green

The artificially fertilized seeds were sown together with several seeds of both original stocks, and the most vigorous examples were chosen for the reciprocal crossing. There were fertilized:

— The hybrids with the pollen of AB

— The hybrids with the pollen of ab

— AB with the pollen of the hybrids.

— ab with the pollen of the hybrids.

For each of these 4 experiments the whole of the flowers on 3 plants were fertilized. If the above theory be correct, there must be developed on the hybrids egg and pollen cells of the forms AB,Ab,aB, ab, and there would be combined:

— The egg cells AB, Ab, aB, ab with the pollen cells AB.

— The egg cells AB, Ab, aB, ab with the pollen cells ab.

— The egg cells AB with the pollen cells AB, Ab, aB, and ab.

— The egg cells ab with the pollen cells AB, Ab, aB, and ab.

— From each of these experiments there could then result only the following forms:

— AB, ABb, AaB, AaBb

— AaBb, Aab, aBb, ab

— AB, ABb, AaB, AaBb

— AaBb, Aab, aBb, ab

If, furthermore, the several forms of the egg and pollen cells of the hybrids were produced on an average in equal numbers, then in each experiment the said 4 combinations should stand in the same ratio to each other. A perfect agreement in the numerical relations was, however, not to be expected since in each fertilization, even in normal cases, some egg cells remain undeveloped or subsequently die, and many even of the well-formed seeds fail to germinate when sown. The above assumption is also limited in so far that while it demands the formation of an equal number of the various sorts of egg and pollen cells, it does not require that this should apply to each separate hybrid with mathematical exactness.

The *first and second* experiments had primarily the object of proving the composition of the hybrid egg cells, while the *third and fourth* experiments were to decide that of the pollen cells. As is shown by the above demonstration the first and third experiments and the second and fourth experiments should produce precisely the same combinations, and even in the second year the result should be partially visible in the form and color of the artificially fertilized seed. In the first and third experiments the dominant characters of form and color, A and B, appear in each union, and are also partly constant and partly in hybrid union with the recessive characters a and b, for which reason they must impress their peculiarity upon the whole of the seeds.

All seeds should therefore appear round and yellow, if the theory be justified. In the second and fourth experiments, on the other hand, one union is hybrid in form and in color, and consequently the seeds are round and yellow; another is hybrid in form, but constant in the recessive character of color, whence the seeds are round and green; the third is constant in the recessive character of form but hybrid in color, consequently the seeds are wrinkled and yellow; the fourth is constant in both recessive characters, so that the seeds are wrinkled and green. In both these experiments there were consequently four sorts of seed to be expected; namely, round and yellow, round and green, wrinkled and yellow, wrinkled and green.

The crop fulfilled these expectations perfectly. There were obtained in the

- — 1st Experiment: 98 exclusively round yellow seeds;
- — 3rd Experiment: 94 exclusively round yellow seeds.
- — 2nd Experiment: 31 round and yellow, 26 round and green, 27 wrinkled and yellow, 26 wrinkled and green seeds.>
- — 4th Experiment: 24 round and yellow, 25 round and green, 22 wrinkled and yellow, 27 wrinkled and green seeds.

There could scarcely be now any doubt of the success of the experiment; the next generation must afford the final proof. From the seed sown there resulted for the first experiment 90 plants, and for the third 87 plants which fruited: these yielded for the

1st Exp.	3rd Exp.	
20	25	round yellow seeds ....AB
23	19	round yellow and green seeds ....ABb
25	22	round and wrinkled yellow seeds ....AaB
22	21	round and wrinkled green and yellow seeds.AaBb

In the second and fourth experiments the round and yellow seeds yielded plants with round and wrinkled yellow and green seeds .... AaBb.

From the round green seeds plants resulted with round and wrinkled green seeds,.......................Aab.

The wrinkled yellow seeds gave plants with wrinkled yellow and green seeds, aBb.

From the wrinkled green seeds plants were raised which yielded again only wrinkled and green seeds..........................ab.

Although in these two experiments likewise some seeds did not germinate, the figures arrived at already in the previous year were not affected thereby, since each kind of seed gave plants which, as regards their seed, were like each other and different from the others. There resulted therefore from the

2nd. Exp.	4th Exp.	
31	24	plants of the form AaBb
26	25	plants of the form Aab
27	22	plants of the form aBb
26	27	plants of the form ab

In all the experiments, therefore, there appeared all the forms which the proposed theory demands, and they came in nearly equal numbers.

In a further experiment the characters of *flower-color and length of stem* were experimented upon, and selection was so made that in the third year of the experiment each character ought to appear in *half* of all the plants if the above theory were correct. A,B,a,b serve again as indicating the various characters.

—A, violet-red flowers;

—a, white flowers;

—B, axis long;

—A, axis short.

The form Ab was fertilized with ab, which produced the hybrid Aab. Furthermore, aB was also fertilized with ab, whence the hybrid aBb. In the second year, for further fertilization, the hybrid Aab was used as seed parent, and hybrid aBb as pollen parent.

- Seed parent: Aab;
- Pollen parent: aBb;
- Possible egg cells: Ab,ab;
- Pollen cells: aB, ab.

From the fertilization between the possible egg and pollen cells four combinations should result, namely.:

$$AaBb + aBb + Aab + ab$$

From this it is perceived that, according to the above theory, in the third year of the experiment out of all the plants,

half should	have violet-red flowers (Aa)	Classes 1, 3
" "	" white flowers (a)	" 2, 4
" "	" a long axis (Bb)	" 1, 2
" "	" a short axis (b)	" 3, 4

From 45 fertilizations of the second year 187 seeds resulted, of which only 166 reached the flowering stage in the third year. Among these the separate classes appeared in the numbers following:

Class	Flower colour	Stem	
1	violet-red	long	47 times
2	white	long	40 "
3	violet-red	short	38 "
4	white	short	41 "

There subsequently appeared,

— the violet-red flower color (Aa) in 85 plants,

— the white flower-color (a) in 81 plants,

— the long stem (Bb) in 87 plants,

— the short stem (b) in 79 plants.

The theory adduced is therefore satisfactorily confirmed in this experiment also. For the characters of *form of pod, color of pod, and position of flowers*, experiments were also made on a small scale and results obtained in perfect agreement. All combinations, which were possible through the union of the differentiating characters duly appeared, and in nearly equal numbers.

Experimentally, therefore, the theory is confirmed that the pea hybrids form egg and pollen cells which, in their constitution, represent in equal numbers all constant forms which result from the combination of the characters united in fertilization.

The difference of the forms among the progeny of the hybrids, as well as the respective ratios of the numbers in which they are observed, find a sufficient explanation in the principle above deduced. The simplest case is afforded by the developmental series of *each pair of differentiating characters.* This series is represented by the expression A+2Aa+a, in which A and a signify the forms with constant differentiating characters, and Aa the hybrid form of both. It includes in 3 different classes 4 individuals. In the formation of these, pollen and egg cells of the form A and a take part on the average equally in the fertilization; hence each form [occurs] twice, since four individuals are formed. There participate consequently in the fertilization

The pollen cells: A+A+a+a,

The egg cells A+A+a+a.

It remains, therefore, purely a matter of chance which of the two sorts of pollen will become united with each separate egg cell. According, however, to the law of probability, it will always happen, on the average of many cases, that each pollen form A and a will unite equally often with each egg cell form A and a, consequently one of the two pollen cells A in the fertilization will meet with the egg cell A and the other with the egg cell a, and so likewise one pollen cell a will unite with an egg cell A, and the other with the egg cell a.

The result of the fertilization may be made clear by putting the signs for the conjoined egg and pollen cells in the form of fractions, those for the pollen cells above and those for the egg cells below the line.

In the first and fourth term the egg and pollen cells are of like kind, consequently the product of their union must be constant, namely A and a; in the second and third, on the other hand, there again results a union of the two differentiating characters of the stocks, consequently the forms resulting from these fertilizations are identical with those of the hybrid from which they sprang. *There occurs accordingly a repeated hybridization.* This explains the striking fact that the hybrids are able to produce, besides the two parental forms, offspring which are like themselves A/a and a/A both give the same union Aa, since, as already remarked above, it makes no difference in the result of fertilization to which of the two characters the pollen or egg cells belong. We may write then A/A + A/a + a/A + a/a = A + 2 Aa + a.

This represents the *average* result of the self-fertilization of the hybrids when two differentiating characters are united in them. In individual flowers and in individual plants, however, the ratios in which the forms of the series are produced may suffer not inconsiderable fluctuations. Apart from the fact that the numbers in which both sorts of egg cells occur in the seed vessels can only be regarded as equal on the average, it remains purely a matter of chance which of the two sorts of pollen may fertilize each separate egg cell.

For this reason the separate values must necessarily be subject to fluctuations, and there are even extreme cases possible, as were described earlier in connection with the experiments on the forms of the seed and the color of the albumen . The true ratios of the numbers can only be ascertained by an average deduced from the sum of as many single values as possible; the greater the number the more are merely chance effects eliminated.

The developmental series for hybrids in which two kinds of differentiating characters are united contains among 16 individuals 9 different forms, AB + Ab + aB + ab + 2ABb + 2aBb + 2AaB + 2Aab + 4AaBb. Between the differentiating characters of the original stocks A,a and B, 4 constant combinations are possible, and consequently the hybrids produce the corresponding 4 forms of egg and pollen cells: AB, Ab,aB,ab, and each of these will on the average figure 4 times in the fertilization, since 16 individuals are included in the series. Therefore, the participators in the fertilization are

Pollen cells:

AB+AB+AB+AB+Ab+Ab+Ab+Ab+aB+aB+aB+aB+ab+ab+ab+ab.

Egg cells:

AB+AB+AB+AB+Ab+Ab+Ab+Ab+aB+aB+aB+aB+ab+ab+ab+ab.

In the process of fertilization each pollen form unites on an average equally often with each egg cell form, so that each of the four pollen cells AB unites once with one of the forms of egg cell AB,Ab aBab. In precisely the same way the rest of the pollen cells of the forms AbaBab unite with all the other egg cells.

10

Research and Development of New Crops

In earlier times, plant improvements were made through a simple selection process based on keen individual observations, in our modern era such improvements are often the result of close cooperation between scientists of many different disciplines and between many different government, industrial and academic institutions. There are still opportunities for a single person or organisation to develop and introduce new plant materials, especially for fruits, vegetables, ornamentals, or traditional field crops, but when it comes to developing wild plant species into new industrial crops, it requires a scenario that depends on tight coordination and cooperation between many organisational entities for success.

The Northern Regional Research Center (NRRC) and cooperating laboratories of the Agricultural Research Service (ARS) of the U.S. Department of Agriculture (USDA) and many Land Grant Universities have accumulated 30 years of history and experience in research and development of new industrial crops. Only recently have scientists and engineers begun to see the fruits of their labours in the form of commercial successes. Millions of dollars and tens of thousands of scientist-hours had to be spent before some of these plant species could be grown by farmers for a potential profit. It required input from botanists, geneticists, plant breeders, agronomists, agricultural engineers, plant pathologists, chemists, chemical engineers, animal nutritionists, economists, administrative and production managers, farmers, labourers, company presidents, and users.

The spectrum of activities from initial research through commercialisation includes at least the following phases: germplasm collection, evaluation and maintenance; germplasm improvement through genetics and breeding; crop development and production technology, processing and product development; product and by-products evaluation; production scale-up; technology transfer; commercial production and utilisation. Not included in the research and development (R&D) spectrum is the determination of the need for a new specific crop or new product derived therefrom. Such a need is often established as the result of national or regional considerations, rather than local or state-based issues. They include: dependence on imports for critical and strategic materials, imbalance between imports and exports, surpluses of traditional crops, need for crops that demand less water, present or expected world shortage of a particular product, and creation of new options for farmers and chemical companies for an improved economy.

Maintenance of Germplasm

International rules need to be observed and honoured. Quarantine regulations have to be followed. In some situations, wild and improved germplasm is well represented and maintained at the National Seed Storage

Laboratory or at the Plant Introduction Stations when such germplasm is appropriately deposited for safe keeping. For other new crops that have received reasonable levels of research attention, little attempt has been made to preserve wild and improved germplasm and subsequently much of it has been lost or can only be retraced with difficulty. With increased international attention to germplasm preservation in general and to potentially useful germplasm in particular, it behooves the Government and individual researchers to protect, evaluate, and share new crops germplasm to the fullest extent possible.

From genetic and breeding research through crop development to processing and product development, the sum total of resources required, both in terms of money and of needed technical expertise, makes it almost impossible for any one state or company to shoulder the burden. It requires a long-term commitment as well, often in the order of 15 years or more, before there is any hope for meaningful levels of new crop production.

Past experience has shown close cooperation between government, academic and private laboratories to be the most beneficial approach to the R&D process. Commitments to funding, staffing and facilities can be shared as agreed and required. Early entry of the private sector in the R&D process has the added advantage of early technology transfer and desire to succeed in the commercialisation phase.

Transfering Technology

Guiding a new field crop into commercial production is a major operation. It requires a well-orchestrated and timed scenario. Farmers need to be informed about the peculiarities of the new plant in regard to production and harvesting. Since there is no open commodity market for the crop, the farmer needs a good contract up front. This establishes movement of the harvested crop and establishes a legal commitment between buyer and grower. Planting seed and extension service need to be provided. A processing facility needs to be contracted before the crop gets planted, so that the crop can be processed at the appropriate time into the desired products. For industrial crops, rarely will the processor be the end user of these products.

More likely, they will be sold to chemical companies that will use the product(s) as intermediates in the manufacturing of its finished goods. Processors and end users also need extensive attention in terms of technology transfer. Rarely are producer, processor and end users found in the same state or even in the same region. This fact alone is sufficient to point to the Federal Government as the prime organisation for technology transfer for the introduction of new crops.

Improvement of Commercial Production

Of the whole development spectrum the most difficult segment has always been the final step, achieving sustained commercial production. That is the crucial step when all the puzzle's pieces have to fall in place. Unfortunately, that is also the time when the private sector often comes into play for the first time. Farmers need to be convinced of forsaking maize, wheat, rice, cotton, or other familiar crops for some new crop and need to be educated about its production. Processors need to be found for toll crushing or other processing contracts. Product end users need to be talked into playing a lead role in the contracting process with farmers and processors, and into giving up familiar feedstocks for one that is not guaranteed to come in when planned and probably at a higher price for several years to come.

It is this setting that requires strong backing and cooperation from the Federal Government. However, it is not a setting where diversion of limited research dollars to this effort is sufficient to accomplish the multitude of tasks at hand. It requires significant input from scientists and engineers, from extension agents, and from strategists and other management experts. Above all, it probably needs financial incentive for a bridging period, which may be anywhere from 2 to 5 years, depending on the crop or product. Such assistance could be in the form of tax incentives, loan guarantees, direct subsidies, joint ventures, or perhaps just a commitment to continue an interactive R&D program to solve problems as they arise.

New Industrial Crops

In US, the Northern Regional Research Center initiated major R&D programs to develop domestic sources of long-chain fatty acids (crambe, meadowfoam, rapeseed), medium-chain fatty acids (cuphea, Apiaceae), hydroxy acids (lesquerella), epoxy acids (vernonia, stokes' aster), liquid wax esters (jojoba), and paper fiber (kenaf). It also assisted in efforts by others to introduce new species into American agriculture. Brief descriptions of the status of each new crop follows.

Crambe (**Crambe abyssinica**)

This source of erucic acid, now mainly imported in the form of rapeseed oil, is ready for commercialisation but has not yet become successful. Oil can be used for many product lines over and above traditional use. Meal has been cleared by FDA for use in beef cattle feed. Crambe could benefit from one or a few good champions in the private sector to further its cause. Certified planting seed is available.

Meadowfoam (**Limnanthes alba**)

Small areas of this source of C20 and C22 fatty acids are produced regularly in Oregon under the auspices of the Oregon Meadowfoam Growers Association. Oil markets are gradually developing. There appears a strong interest in Europe to use meadowfoam as a local new crop for fatty acid production. Strong research programs are still active at Oregon State University and NRRC for crop and product development.

Rapeseed (**Brassica napus, B. campestris**)

With the advent of canola-type rapeseeds, production of traditional high-erucic acid rapeseed has waned rapidly worldwide. Up to 12 years ago, USDA and Oregon State University cooperated to create new lines of high-erucic, low-glucosinolate rapeseed for production in the USA.

Cuphea (**Cuphea** *spp.*)

This program serves as an excellent example of international cooperation between companies, governments and universities. Resources are pooled and annual conferences are held to review progress and to plan new work. Although from a chemical point of view no better genus of plants is known as a source of C8, C10, and C12 fatty acids, all species are difficult to domesticate due to indeterminate flowering, seed shattering, and the presence of sticky hairs covering the plants. Research is being carried out at many locations. Financial commitments have been made by all cooperator sectors.

Apiaceae (formerly: Umbelliferae)

This plant family contains many species that produce seed oils with high levels of petroselenic acid (delta 6-monounsaturated C18 acid). This acid can be split oxidatively into C12 lauric acid and C6 dibasic adipic acid. The latter is used commercially for nylon production. Common fennel is one species among several being evaluated for agronomic potential, but no integrated and well-organised effort is known to exist.

Lesquerella (Lesquerella fendleri)

This genus contains various species that produce several unsaturated hydroxy fatty acids, which are similar but not equal to the C18 ricinoleic acid from castorbean oil. Genetic and agronomic work is carried out by the ARS in Arizona, and the NRRC is evaluating the oil for grease formulations and other utilisation opportunities. The oil may become prominent in coatings, lubricants and plastics technologies if given a chance.

Vernonia (Vernonia galamensis)

Initial work in the 1960s was carried out with *Vernonia anthelmintica* but was switched to *V. galamensis* in the 1970s. It now appears that epoxy acid-containing vernonias are not suitable for American agriculture due to the requirement of short day length for flowering and seed setting. ARS is assisting several African nations near the Equator in their efforts for crop development. Several laboratories in the U.S. and Europe are evaluating the oil in coatings and adhesives formulations. This potential crop species appears to be heading for success.

Stokes' Aster (Stokesia laevis)

This native American species contains an epoxy acid seed

oil almost identical to that of *Vernonia*. It has received scattered production research attention over the years, but no concentrated, organised effort is anticipated at this time.

Jojoba (**Simmondsia chinensis**)

Considering the drawback in production start-up costs due to the perennial nature of jojoba, this new crop can be considered a success story. The U.S. alone has now about 40,000 acres under cultivation, supported by an active growers association and a marketing cooperative. Sizable areas are now cultivated also in Mexico, South America, Australia, Israel, and India. Many other countries with warm dry climates are considering jojoba production. Oil production is increasing, but demand is expected to stay even with or to outpace production. Production and utilisation research is continuing with excellent cooperation between the various sectors.

Kenaf (**Hibiscus cannabinus**)

This annual species is a source of paper fiber and has become another success story. After 20 years of excellent production and processing research and 10 years of industrial evaluation, a commitment to invest over $300 million for a new pulp and paper mill has been reached by the private sector. The mill is being built in Texas and is based on 100% kenaf feedstock, which is to be produced on approximately 10,000 ha by farmers in the vicinity of the mill. This effort is only a beginning to reduce our $4 billion annual trade deficit because of paper import. It also creates new business and labor opportunities. A concentrated technology transfer effort by USDA contributed greatly to this success.

Industrial Perspectives of Research and Development

Agriculture is America's largest industry and, for a variety of reasons, a strategically critical one. Consequently, agriculture receives much government intervention. Quota systems limit the production of certain crops, while federal commodity programs place price floors under certain crops and restrain their production. Nonetheless, as a nation, American are moving progressively toward a more market-driven agricultural industry, while concurrently edging the rest of the world down the same path. This means agriculture will increasingly need to focus on producing crops to meet the current market demands which provide the best opportunity, for profit. Consequently, in the face of a changing and increasingly competitive world environment, it seems tha t agriculture will necessarily become more like other large industries.

Generally speaking, industries which have an overconcentration or overdependency upon a narrow line of unchanging products are quite vulnerable. Changes in market demand or competitive activities can lead to reduced sales and unprofitability. In a truly free market situation, such scenarios can lead to the virtual demise of an industry. In order to maintain long term growth and profitability, any large industry must necessarily diversify, through the development of new products. This is accomplished through the process of innovation—a process by which new product ideas or concepts are developed and commercialised.

Conceptually, agriculture needs increased product options-alternative crops which address different market needs and can be produced on an annual basis. Such increased options can enable American farmers to maintain production and maximise profits by responding to changing market conditions. A national task force recently addressed this issue' and concluded that diversification of American agriculture and forestry must become a national priority. Specifically, the New Farm and Forest Products Task Force concluded that "significant opportunities exist for new farm and forest products to meet real market needs—particularly in industrial, non-food application areas."

The development of new crops is fundamentally the practice of innovation in agriculture. In its essence, innovation may be thought of as a process of adding value to ideas. It is, by any measure in any industry or product category, a complex, risky and lengthy process. Although the actual process of innovation is quite complicated and difficult, it is useful to think of it as a linear, three-step one. Under this simplistic definition, we can think of the first step as that of invention, or discovery.

Applying this step to new crop innovation, the discovery would likely occur in a governmental or university laboratory, or an agricultural experiment station. A scientist or researcher might identify a prospective new crop candidate by discovering properties in a wild plant which might be of commercial value. Illustrations of such discoveries might include kenaf, guayule, meadowfoam, or even cuphea.

Alternatively, the researcher may have conceived of a way to modify an existing crop, perhaps by genetically changing the rapeseed to produce a different oil with new chemical market potential. In this first stage, the challenge is to establish plausibility for the new crop. Those crops which offer sufficient promise may progress to the next stage, that of translation of the conceptual crop into a real one. In this stage, the feasibility of growing the crop and delivering the right product must be demonstrated. This is a lengthy and far more expensive part of the process—wherein the true commercial potential must be determined. The economics and technology must be right, and the market must be ready.

It should seem obvious that in this stage, there must be a close working partnership between the original researchers, the industrial users of the prospective crop product, and everyone in between. In other words, there must be close cooperation between the technologists and the marketers. Producers must also be involved to demonstrate the feasibility of crop production in practical terms. Generally, the need is defined for a public and private sector, cooperative partnership in some form. The needed levels of cooperation are, of course, difficult to obtain in this stage because of the uncertainties, the risks, and the costs involved.

Obviously, the third stage in this process is that of full scale commercialisation. This is the moment of truth when large scale production is attempted, and the new crop is integrated into the whole production, distribution, and usage system.

To commercially succeed, everything must work right. The producer must be able to grow the crop and sell it for a profit. Industry must be able to buy the crop, add value to it through processing, and sell the resulting products at a profit. Finally, the consumer or final user must accept the finished product as one which meets a need and has sufficient value to justify its cost. All along the chain, additional handlers, shippers and distributors must also recover the costs of their service plus an appropriate profit.

As the complexity of the process suggests—even under this simple model, there are many more ways to fail than there are to succeed. Unfortunately, we do not have a sufficient base of data in agricultural developments to meaningfully estimate the probability of any new crop concept successfully progressing to market. However, based upon a considerable base of data in other industries throughout the world, an average of only one to five per cent of formally evaluated ideas, concepts or inventions ever make it to the marketplace.

This does not mean the ones which become commercial realities are big winners. Indeed, the data suggest that only one in one thousand become truly big winners. The odds might seem better than in a state lottery, but the costs of buying a ticket are decidedly much higher. It is important to keep in mind that, in a private sector context, the successes must be big enough to pay back not only the cost of that specific development, but the costs of all the preceding failures as well.

On top of such sobering odds against success, innovators in any field are facing long timetables. On average, it takes seven to ten years to move an idea successfully into the marketplace, with agricultural innovations generally taking much longer than that. Such long timescales mean that it is quite difficult for the private sector to generate the financial rates of return needed to justify the investments involved, particularly in the early stages of development.

Although it restates the obvious, it is useful to reflect on the fact that the level of financial returns needed by the private sector are largely determined by the cost of money. Like any other commodity, money has a cost, and a time-value. Even when interest rates are low, the cost of money is high if it will be a long time before it is paid back. In agricultural developments, the situation is even more complicated, with the financial returns from new crops dispersed between producers, processors and consumers.

Improving Commercial Technologies

Many people think that new ideas flow easily from the research bench to the marketplace. Nothing could be further from the truth. Scientific laboratories and libraries are overflowing with alternative techniques for producing, processing and marketing agricultural products. Some of these techniques are ready for adoption and others are not. Sorting out the most promising ones is difficult and time consuming. An estimated 90% of the cost of research and development is development.

Development—demonstrating the usefulness of a promising technology-includes growing and harvesting technology, raw materials handling systems, processing and extraction technology, market identification and promotion, business planning, and formation of grower cooperatives and associations. The purpose of development is two-fold: to prepare the farm sector to grow and sell a material for new use and to prepare industry to buy, process, and market the product. The weakest link in the U.S. economy is transferring promising technologies from "Idea Generation" to "Interim Manufacturing" stages. There is little strategic planning or investment in phasing out obsolete technologies and developing new ones that satisfy changing markets.

Public supported agriculture research and education have always had a close working relationship with producers of food, fiber, and forest products. Most of the effort has been on production-increasing yield per acre, animal unit, or hours of labor. That was fine when the

United States fed, clothed, and sheltered its own citizens and exports were continually expanding. The value of agriculture exports is forecast to climb $4.1 billion from 1987 (USDA projections). Increased sales are expected in grains, soybeans, and horticulture products. This is good news for farmers, but it is unlikely that the 1981 record levels can be attained in the near future. Primary reasons include: the relative value of the U.S. Dollar has fallen significantly as compared to the Japanese yen and many Western European currencies, but has changed little in comparison to major agriculture exporting nations like Canada, Brazil, and Australia; expanded food producing capacity in countries like China and India; import restrictions in Asian and West European markets; aggressive targeting of U.S. markets by many countries; and lack of foreign exchange in less developed countries-where the largest growth in population is occurring.

One possible solution is to put promising technologies for farm products in the hands of private investors to provide these markets. Every other country is anxious to target U.S. markets. Why shouldn't U.S. scientists, technical specialists, processors, material handlers, venture capitalists, and marketers put their heads together and find a way to tap one of the best markets in the world. This doesn't mean that export markets shouldn't be aggressively pursued, but let's also explore opportunities within our own borders.

In the United States, private and public institutions have often mistrusted each other. An entrepreneur sees public institutions as tax collectors or regulators, unaccountable for the bottom line. A civil servant sees business as having a short-time horizon, focusing primarily on making money, and having inadequate concern for broad public interests such as environmental and social issues. These stereotype images are common enough to prevent needed bridge building.

Many research programs are funded by universities and the Federal Government because payoff is too long term and uncertain for the private sector. Commercialisation of promising technologies has similar incentive problems, but is not quite as elusive. Therefore a partnership is the answer. Joint private/public ventures are needed because very little happens without a combined effort. A joint partnership helps share the risk, especially in the early stages. The private sector has a key role to play in the process because people experienced in buying and selling goods are best able to identify the market. The private sector also should provide a product champion, that is, a firm or individual that believes in the ultimate commercial success of a particular product. A project should be based on solid analysis, but the success of any given venture most often depends on someone who strongly believes in the final outcome. Industry members of a commercialisation team have the knowledge and experience to identify the best market opportunities and the most useful knowledge about technologies to meet changing competitive conditions. They also understand why private sector initiative provides the most flexible and efficient mechanism for producing, processing, and marketing agricultural products.

Government/academic partnerships can provide well-trained researchers and technology transfer specialists with a long-term view; a network of State and Federal Government contacts who can reduce regulatory red tape; and provide needed seed money. They also can bring production, processing, and marketing sectors together. Often, not enough incentive is present in any one sector to adopt promising technologies, but together the incentives can often be sufficient.

A commercially successful and profitable enterprise does not always require a joint effort. Private sector initiatives frequently create new markets for farmers. In many instances, however, the barrier to commercialisation is management and availability of resources with which to link agriculture production to the processing and marketing sector. When appropriate, a catalyst organisation can provide that linkage by actively organising and promoting the final stage of product development with relatively small public involvement. Time is of the essence for developing promising ideas. If the appropriate moment is not seized, the advantage of early adoption will occur elsewhere, most often by a foreign competitor.

Demonstration Projects

In recent years the Special Projects Unit in the Cooperative State Research Service has been bridging the gap between research results and commercialisation with demonstration projects. These projects are short term, usually 3 to 5 years, with public/private partnerships established through USDA cooperative agreements. Capitalising on decades of public research-generally by the Agricultural Research Service or State Agricultural Experiment Stations—the demonstration projects are designed to build, within the private sector, a base in which to sustain new agricultural industries. The kenaf, guayule, and hybrid striped bass projects demonstrate the success of this approach.

Kenaf project

Kenaf (*Hibiscus cannabinus*) is an annual, nonwood fiber crop that grows from seedling to 10 feet in less than 5 months and can be transformed from seedling to newsprint in less than 7 months. It grows across the southern tier of the United States. Newsprint is normally made from wood such as northern spruce and southern pine. The United States imports about 60 percent of its newsprint supply from Canada at an annual cost of $3 to $4 billion. The Kenaf Demonstration Project began in March 1986 with a cooperative agreement between USDA and Kenaf International—a venture capital firm and the product champion for kenaf. This agreement marked the reopening of research and development for kenaf in USDA after an 8-year hiatus.

The project's principle objective was to gain acceptance for kenaf as a source fiber for the manufacture of newsprint. To accomplish this objective, a Kenaf Task Force was established involving participation of a venture-capital company, private laboratories, processing and pulping plants, equipment manufactures, and the American Newspaper Publisher's Association-a prime mover in the search for a substitute newsprint fiber.

Guayule project

The United States is totally dependent on imported natural rubber—a commodity of critical importance to the Nation's defense position and the economy. Approximately 750,000 metric tons are imported at an annual cost of $1 billion. Natural rubber has performance specifications not available in its synthetic alternative, and is preferred in specifications that require elasticity, resilience, tackiness, and low heat buildup. Natural rubber constitutes 30 percent of the total domestic rubber market.

Of some 2,000 plant species known to contain rubber, only a few have ever produced sufficient quantities for commercial use. The two most common sources are the rubber tree (*Hevea brasiliensis*) grown principally in Southeast Asia (where 90% of the world's supply of natural rubber originates) and guayule, (*Parthenium argentatum*) native to semi-arid regions of North America. In contrast to the rubber tree, guayule is a small desert shrub averaging 3 feet tall.

Under a 27-month guayule domestic rubber project signed in 1986 with USDA, the Defense Department (DoD) is providing $11 million to build and operate a prototype processing plant in Arizona that will process 275 acres of guayule shrub into 50 tons of natural rubber, 100 tons of resin and low molecular weight rubber, and 1,600 tons of plant residue. Also cooperating on this project are the Firestone Tire and Rubber Company, Gila River Indian Community, and four southwestern land-grant universities-located in Arizona, California, Texas, and New Mexico. Over this same period, USDA plans to invest about $4.5 million in guyaule research and development.

With today's technology and current prices for national rubber, a domestic industry is not economically viable. An economic model indicates a farmer growing guayule would have positive cash flow in 1995, if the following technological improvements could be realised: increasing yields of rubber and byproducts 21/2 times, developing direct seeding capability, establishing markets for byproducts, and reducing production, harvesting and processing costs by one-third.

Hybrid Striped Bass (HSB) project

A field trial demonstration project for raising HSB in farm ponds began operation in the fall of 1986 at Walnut Point Farm near Chestertown on Maryland's Eastern Shore. HSB is a cross between the female striped bass, or rockfish, (*Monroe saxatilis*) and the male white bass (*M. chrysops*). It has the same flavor and general appearance as a striped bass but is more disease-resistant and faster growing than either of its parents.

The 3-year development phase includes construction of 8.5 acre fingerling (young fish) production ponds, four 2.5 acre growout ponds; six water supply wells, and a combination hatchery, storage, and Extension education building the first year; it will include two additional 5-acre growout ponds the second year; at maturity 24 surface acres of water, 4 dedicated to fingering production and 20 dedicated to the production of market-size fish.

The increasing demand and decreasing supply of fish and fish products have resulted in a dramatic jump in domestic product cost and foreign imports. The U.S. seafood consumer now pays 64 cents of every seafood dollar to foreign suppliers. These imports amounted to over $5.0 billion in 1986, an historic record, and contribute to the Nation's balance of payments deficit. Since the harvest from our wild fisheries is not growing appreciably, the increasing consumer demand for fish can only be met by foreign suppliers or by a strong domestic agriculture industry. Through joint ventures, U.S. aquaculture can develop to offset these escalating imports.

The market potential for HSB is promising. Because of the moratorium on striped bass fishing in the Cheaspeake Bay, there is an immediate market shortfall of 14 million lbs. Private and government marketing experts estimate a beginning market of 54 million lbs-equal to that for trout. At that production level, producers' gross income would be about $182 million. The margin between $1.00-$1.50 estimated per pound costs and current New York price of $4.50-$5.00 per pound appears adequate to attract private investment, but there are technical and market uncertainties.

Strategic Planning for Development

The U.S. government has spent more than $200 billion in the decade 1978-87 on domestic programs related to surplus agricultural commodities. These multi-billion dollar outlays in such things as commodity price supports, export programs and farm credit bailouts, together with record numbers of farm bankruptcies, suggest a policy-stimulated over-concentration of U.S. agriculture on a few primary commodities.

The biotechnology industry somewhat paradoxically has contributed to the problem as well-as agricultural production techniques become more efficient, commodities surpluses increase further. One biotechnological product, for example, enhances livestock productivity, thereby reducing the quantity of feed required to market animal products. The projected decrease in feed demand would add an estimated 4.6 to 10.3 million acres of excess crop production to U.S. agriculture, expanding excess crop production capacity-already estimated at 40 million acres-by 9 to 25%. Yet biotechnology could be applied to improve new crops for profitable production of imported industrial raw materials instead of producing more surplus commodities.

The use of biotechnological methods in agriculture is somewhat analogous to the employment of robotics in manufacturing. While improved manufacturing efficiency may be large, total national health and economy could experience severe impacts from labor displacement and insufficient alternatives for productive reemployment. The solution, for both industry and agriculture, lies in diverting resources to new products and services with elastic markets. In much the same way that displaced workers have been retrained, sufficient investment to develop new crops can help convert the excess U.S. agricultural production capacity to serve society better.

While agriculture is the largest single contributor to the U.S. economy and arguably a very efficient industry, given the chronic nature of surpluses, scientific production of only the major crops may no longer be enough. Farming, which constituted only 11.3 percent of the agricultural sector contribution to GNP in 1982, needs a boost in economic return. Also, new industrial crops and products could cycle within the domestic economy as value-added commodities, generating GNP at rates equal to or greater than GNP generation by subsidised commodities currently produced in surplus and exported with little, or no, modification.

Today, consumer needs that could be met through the utilisation of U.S. agricultural resources include plastics, cosmetics, gasoline and lubricant additives, insecticides, medicines, paper products, super absorbent polymers, and foods low in fat and cholesterol. In addition to substantial economic benefits, new crop development could address a wide variety of environmental concerns. For example, health hazards could be reduced by using vegetable oil to replace petroleum-based plasticisers in industry. Plant-derived lubricants that substitute for sperm whale oil could be developed. Ethanol from biomass could be used to a greater extent as an efficient motor fuel in no-lead gasolines to reduce urban lead levels and control atmospheric carbon levels. .

Moreover, many strategic and essential substances or plant-derived substitutes could be grown in the U.S. Where natural rubber, castor oil, and sperm whale oil must be stockpiled for defense purposes, domestic production of new crop candidates-such as guayule (*Parthenium argentatum* G.) to supply natural rubber, and jojoba [*Simmondsia chinensis* (Link) Schneider] as a source of liquid wax esters-would help insure availability of these strategic materials. Furthermore, new crops could supply essential industrial materials-waxes, resins, gums, newsprint, lubricants, and alternative sources of hydrocarbons used in manufacturing numerous synthetics.

However, the greatest opportunity for new agricultural products lies in applications which do not substitute for or compete with current agricultural products, but do create totally new markets. Meadowfoam (*Limnanthes* spp. R. Br.), a new source of unique long-chain fatty acids, is one example. Indeed, increased efficiency of producing resources, such as the use of potentially lower cost high fructose sweeteners from corn (*Zea mays* L.) to replace cane (*Saccharum officinarum* L.) and beet [*Beta vulgaris (saccharifera)* L.] sugars, could benefit society (even though corn also is subsidized). At the same time, such changes could reduce dramatically sugar crop acreage and spell disaster for cane and beet sugar growers.

Thus, strategic planning and a national-level long-term commitment to buffer the effects of crop acreage displacement by introducing new replacement crops clearly are justified. Soybean introduction and commercialisation on 40 to 60 million acres of corn, wheat [*Triticum vulgare* (*aestivum* L.)]., and cotton (*Gossypium hirsutum* L.) land in the U.S. is illustrative in this respect. Though interest in soybean as an oilseed crop in the U.S. was very limited in the early 1900s, the crisis of domestic vegetable oil shortages during WWII drove up the price of soybean in 1920 to more than $4.00 per bushel. Both the crisis-born profit potential and the strong, enduring, and accountable voice of W.J. Morse, director of soybean research in the USDA, served to break the new-crop entry barrier and transformed soybean into a "Cinderella" crop.

The history of soybean production has a threefold significance: it illustrates the economic potential latent in experimental crops; it underscores the role of a strong political voice in directing public resources to crop development; and it points to the crucial necessity of an ongoing, long-term commitment to new-crop research that can dampen future cycles of agricultural overproduction.

While new farm and forest products should not be viewed as quick or complete solutions to the problems of surplus agricultural commodities, there are substantial grounds for optimism. The historical development of two major, long-domesticaled U.S. crops—corn and soybean—provides suggestive models of how plant productivity can be increased to benefit society. During the 20th century, average U.S. yields of maize and soybean increased three-fold, with about half of this increase owing to plant breeding efforts alone. At the same time, the labor required to produce maize in the U.S. in 1986 was approximately one fiftieth of that needed in 1910.

Similar efficiency and even higher rates of yield improvement could be expected as successful alternate crops and products are developed. Yield improvement in particular is a critical first step in making new crops profitable, and promoting positive attitudes and support for sustained research and development. One example is a socioeconomic analysis of costs and benefits of a 14-year investment made by Canada in improved rapeseed (Canola) cultivars (*Brassica napus* L. and *Brassica campestris* L.) and crop development.

The U.S. particularly needs a mechanism to overcome the entry barrier hindering new crop development. In the early stages of crop development, long-term funding is difficult to maintain. Private industry typically is not interested in funding this type of research until the profit potential is apparent. By the same token, unrealistic expectation from new crops is a problem that can result in a defeatist attitude which reduces support for other research projects on new crops. Scientists themselves are often reluctant to risk investing time in new, uncertain crops, preferring the security of an established crop with existing scientific, financial, and political support. The entry barrier to development is a type of new-crop "catch 22:" funding is not provided until profit potential is demonstrated, and this cannot be done without funding the initial research.

Too often new crops are expected to be profitable before necessary developmental research has occurred to increase yield. Thus, economic analyses of yields of wild plants or unadapted alternate crop cultivars have resulted in unimpressive prospects. Historically, major crops—even those domesticated for millennia—have increased in yield by over three-fold only during the 20th century, given sufficient research investment. Therefore estimates of future yield and profit potential—and setting of research investment for a given new crop—appropriately could be based on yields at least 3 times those experienced upon initial domestication or selection. Added indirect benefits from new crops, such as improved pest control through crop rotation, should be measured and credited to the new crop.

Furthermore, dramatic reductions in costs of producing, harvesting, processing, and marketing can be expected with systems research and engineering developments. Finally, new crop profit potential should be measured against that from traditional crops minus crop subsidies and other distorting influences. If past-crop developments provide an index for new-crop yields, they serve as well to emphasise the importance of making long-term commitments. Approximately 200 years elapsed from the time soybean was introduced in North America until it was established as one of the five major field crops.

While biotechnology may offer opportunities for new-crop development at an accelerated pace, traditional plant breeding and agronomic research will be crucial to improvement efforts. Thus, based on historical timetables for crop improvements, a philosophy and mechanism allowing for a minimum 20-year commitment to improvement of a given new crop is needed and is reasonable. Unfortunately, in the U.S., nothing less than a national shortage of a commodity and the concerted long-term efforts of an individual or an organisation to champion a new crop have been necessary to launch successfully an integrated new crop development strategy at the national level, as is illustrated by the histories of soybean and rapeseed development. This crisis-based policy is policy by default, a short-sighted or nonstrategy policy that is haphazard, very expensive, and attentive only to commodities necessitated by the historical moment.

The cost of improving a new crop, based on precedents, can be estimated at $40 to $840 million. The $12 billion cost to the U.S. government for the 1987 corn-related support programs therefore would be equivalent to the estimated total 20-year cost for improvement of 14 to 300 new crops. The $200 billion spent on farm income stabilisation during the decade 1978-1987 would fund the total 20-year costs of improving 238 to 5000 new crops. Though there are not 5000 high-priority plants to develop as crops, the figures dramatise the staggering disproportion in expenditures. And these calculations do not include the economic opportunity losses from not growing profitable new crops on the millions of acres producing surpluses or idled by government payments.

Funding is a necessary condition, but not a guarantee of the successful commercialisation of a given new crop or new product. However, experience suggests that potential crops chosen carefully could be expected, at least together, to return wealth to the nation comparable to the $518 billion earned by soybean in 60 years. Less than 1 million acres each of 50 new crops (symbolically, one per state) could absorb the current domestic excess production capacity. Funding each of 50 new crops at the annual rate of 15 SYs would cost $140 million—approximately 18% of the USDA budget for the Agricultural Research Service (ARS) in 1987, or 7-9% of the sum of ARS and individual state appropriations (1987), or 0.31% of the 1987 farm income stabilisation costs.

New Organisational Entitiy

Granting that the need exists now for new-crop development, and that it will become even more urgent with improved farming efficiency worldwide, the question arises as to how to implement most effectively research and development on new and alternative crops and products. The private sector cannot be expected to undertake the long-term initial investments needed for new-crop development; a steady public support source is fundamental. The well documented unstable nature of public investment in identifying and developing new crops and new products illustrates the critical economic need for strategic planning in U.S. agriculture. It is appropriate to focus national attention on satisfying future consumer and industrial needs from agriculture.

Conditions, events, and societies support modifying the "agriculture-food" stereotype to include expanded production of nonfood industrial raw materials.

An ongoing entity-a foundation, institute, center or association-has been called for, and Erich Bloch Director of the National Science Foundation, has described many elements of change needed to foster a national research strategy to rebuild U.S. economic competitiveness. The need for new crops could be a rallying point but, as Bloch points out, making the proposed institutional change will require "tough management, innovative polices, and vigorous leadership".

The importance of a well-qualified and unbiased organisation involved in objectively choosing and promoting the most promising new crops can be seen from the structural bias in the political funding process. Funding of agricultural research by the public sector overwhelmingly reflects the interests of producers, handlers, and processors of established commodities and technological areas of topical interest (e.g., genetic engineering and biotechnology). The need exists for a forceful and enduring voice for research into new farm and forest commodities.

An entity responsible for new-crop development would provide the means to develop a strategy, to focus new technology on exploiting renewable wealth from plants, and to solve problems that often prevent their domestication. It could attract scientific talent, leverage resources to priority crops, and promote a collective synergy among diverse interest groups. The development of new crops and products could take place largely within existing agricultural research networks.

The track record of crop improvement research by agencies focusing on specific crops has been highly successful to date. The development of the Canadian rapeseed industry was supported by the Saskatchewan Wheat Pool. Improved potato (*Solanum tuberosum* L.) germplasm was produced by the International Potato Center and used by over 80 countries. These examples attest to the progress possible with long-term support focused on particular crop plants. Other International Agricultural Research Centers, and particularly the International Rice Research Institute, have made remarkable progress in crop improvement.

More than 1000 new crop cultivars had been developed at these centers by 1983. The modern rice and wheat cultivars developed at these centers annually yield enough more than the old cultivars would have produced to feed an additional 500 million people. Ironically, while the U.S. has participated in the planning, development, and support of most of these international centers, it has not established any comparable comprehensive interdisciplinary effort for domestic development and commercialisation of new crop-plant resources.

Multidisciplinary needs in new crop development argue for creating a stable, organised entity to foster cooperative efforts. The establishment of entities to coordinate research efforts and to focus on vital, multidisciplinary areas has been identified as essential to U.S. economic health. Recent proposals to develop 80 to 100 science and technology centers at U.S. universities by 1992, to create a "National Institute of Technology", and to establish an "Advanced Civilian Technology Agency" reflect the growing awareness of the need for federal action to increase U.S. economic competitiveness.

International Repercussions

New crops—or old crops which undergo major

improvements—can make massive contributions to economic growth and development. On the international scene, major improvements in both wheat and rice cultivars during the 1960s triggered unprecedented production increases in environments where the advanced technologies were well adapted. So great was their contribution to overcoming hunger that collectively we came to call the use of these new technological packages the green revolution.

It was hoped that Norman Borlang, who is often called the father of the green revolution, would participate in this symposium. As he was unable to be present it seems appropriate to relate the following anecdote. It involves Borlang's new semidwarf wheat, 'Mexipack 65'. The place was what was then West Pakistan; the time, 1968. Norm and other members of our party were examining wheat fields, visiting with farmers and public officials. International Maize and Wheat Improvement Center (CIMMYT) began working on the crop in 1966.

Then CIMMYT's objective was to develop a grain crop that had the advantages of both wheat and rye but the disadvantages of neither. Triticale, the first man-made commercial crop, made its initial appearance in the 1870s. As the product of a cross between wheat and rye it remained little more than a botanical oddity until well into the twentieth century. The crop's development up to that time and since has involved a broad spectrum of scientists in many countries. Since the initial crosses were made by Stephen Wilson in Scotland, triticale has commanded the attention of plant breeders in Germany, USSR, Sweden and in the 1950s, Canada, Spain, Hungary and Japan. The big international push, however, came after the University of Manitoba started a joint program with CIMMYT.

The financing of blue-sky initiatives such as the development of a new commercial crop like triticale is of interest. The Rockefeller Foundation, a cosponsor of CIMMYT, helped launch the effort. Canada's interest stemmed in part from the hope that they might transfer rye's resistance to leaf diseases into their durum-wheat crop. But the difficulties in the then-existing triticales were challenging indeed. The tall late maturing plants lodged, were sensitive to day length, partially sterile, low yielding, and tended to set shriveled seed. The seed was subject to preharvest sprouting, germinated poorly, the quality for milling and baking left much to be desired, and market demand was an uncertain if not unknown quantity.

The breeders had faith that if they could assure a supply of the new crop, somehow that supply would generate its own demand. In 1970—two Canadian organizations—The International Development Research Center (IDRC) and The Canadian International Development Agency (CIDA)—provide CIMMYT with $2.5 million for a 5-year triticale development program? They were gambling that if they were successful they would produce a commercial crop with the yield levels and food quality of bread wheat which also had rye's tolerance of cold stress, moisture stress, and poor acid soils.

Rural hunger and poverty problems were often greatest in many marginal areas where an improved triticale might fit. Canadians believed in CIMMYT and wanted its important work to move forward. But with a surplus of wheat at home it might not be politically expedient for Canada to support CIMMYT's wheat program. Opponents might view such efforts as building up competitive producers thus contributing to decreases in Canadian wheat exports. Triticale was not wheat. It was a different crop. Help to the less advantaged and national pride rather than political dissension could well be associated with a joint Canada-CIMMYT triticale initiative. So the grant was made.

Drawing on its international support, CIMMYT has now sustained a substantial triticale program for more than two decades. Annually this has involved making 1,000 to 2,000 crosses and keeping 30 to 40 million triticale plants under observation. For 20 years the Center has distributed international screening and yield nurseries to more than 70 countries. Yields are being tested in over 60 selected locations worldwide. Thus the developmental effort in the public sector in which CIMMYT is a key actor is of major proportions.

Now, more than 115 years since the crop's discovery, due to cooperation of plant scientists throughout the world triticale has become a commercial crop. CIMMYT's trials indicate that yields now approximate those of wheat and rye in areas where these crops are traditionally grown. Triticale's chief advantages, however, are its capacity to out yield wheat and rye in many marginal crop production areas. Polish triticales, winter types, have spread most rapidly, notably in Europe and the USSR. However, spring triticales now show promise in the drier parts of North Africa, for example. In such environments they often out-compete wheat at the lower range of adaptation and performance just as they frequently do better than barley in the more favorable portions of the ecologies where that crop is grown. It is in such marginal or stressful dry lands where soils are sandy, acid, and phosphorous-deficient that triticale appears destined to make net additions to food availability.

Bibliography

Altman, D.W. and Watanabe, K.N., *Plant biotechnology transfer to developing countries*, Academic Press, Inc., California and London, 1995.

Arakawa, T., Chong, K.X. and Langridge, W.H.R. "Efficacy of a food plant-based oral cholera toxin B subunit vaccine", *Nature Biotechnology*, 1998.

Artsaenko, O., Kettig, B., Fiedler, U., Conrad, U. and Düring, K., "Potato tubers as a biofactory for recombinant antibodies", *Molecular Breeding*, 1998.

Bernardo, R., "A model for marker-assisted selection among single crosses with multiple genetic markers", *Theoretical Applied Genetics*, 1998.

Bourlag, N.E., "Feeding a world of 10 billion people: the miracle ahead", *Plant Tissue Culture and Biotechnology*, 1997.

Briggs, S.P., "Plant genomics: more than food for thought", *Proceedings of the National Academy of Sciences* USA 95: 1986-1988.

Busch, L., Lacy, W.B., Burkhardt, J. and Lacy R.L., "Plants, power, and profit: social, economic, and ethical consequences of the new biotechnologies", Basil Blackwell Inc., Cambridge, Massachussetts, 1991.

Chee, M., Yang, R., Hubbell, E., Berno, A., Huang, X.C., Stern, D., Winkler, J., Lockhart, D.J., Morris, M.S. and Fodor, S.P.A., "Accessing genetic information with high-density DNA arrays", *Science*, 1996.

Lee, M., "Genome projects and gene pools: New germplasm for plant breeding?" *Proceedings of the National Academy of Sciences* USA 95: 2001-2004.

Lemieux, B., Aharoni, A. and Schena, M., "Overview of DNA chip technology", *Molecular Breeding,* 1998.

Marshall, A. and Hodgson, J., "DNA chips: an array of possibilities", *Nature biotechnology,* 1998.

Paterson, A.H. (ed.), "Genome mapping in plants", Academic Press, Inc. and R.G. Landes Co., New York and Austin, 1996.

____________ . (ed.), *Molecular dissection of complex traits,* CRC Press, Inc., Boca Raton and New York, 1997.

Ramsay, G., "DNA chips: state of the art", *Nature Biotechnology,* 1998.

Rifkin, J., *The biotech century,* Victor Gollancz, London, 1998.

Tacker, C.O., Mason, H.S., Losonsky, G., Clements, J.D., Levine, M.M. and Arntzen, C.J. "Immunogenecity in humans of a recombinant bacterial antigen delivered in a transgenic potato", *Nature Medicine,* 1998.

Troyer, A.F., *Breeding widely adapted, popular maize hybrids,* Euphytica, 1996.

Wallace, D.H. and Yan, W., "Plant breeding and whole-system crop phyisiology: improving adaptation, maturity, and yield", *CAB International,* Wallingford, Oxon, 1998.

Index

enhanced over those of castor oil because of the increased chainlength of this new crop oil.

Like most of its cruciferous relatives, *Lesquerella*'s seed meal contains glucosinolates. There are several different glucosinolates identified in *Lesquerella* spp. Many of these can be removed easily from the meal if converted to the volatile isothiocyanates. The amino acid composition of the seed meal has been reported. Its especially high lysine content makes it attractive as a protein supplement.

Jojoba (Simmondsia chinensis)